Conservation of Clay and Chalk Buildings

To
Jenny for patience,
Dave for support
and
Chris for faith

Conservation of Clay and Chalk Buildings

GORDON T. PEARSON

Professional Associate of the
Royal Institution of Chartered Surveyors
Post Graduate Diploma in
Building Conservation
(Architectural Association)

Routledge
Taylor & Francis Group

LONDON AND NEW YORK

First published 1992 by Donhead Publishing Ltd

Published 2015 by Routledge
2 Park Square, Milton Park, Abingdon, Oxon, OX14 4RN
711 Third Avenue, New York, NY 10017, USA

Routledge is an imprint of the Taylor & Francis Group, an informa business

ISBN 978-1-873394-00-7 (hbk)

A CIP catalogue for this book is available from the British Library.

Typeset by Keyboard Services, Luton

Contents

Chapter Three
The qualities of earth walling   **43**

Preface

It was not until the autumn of 1977 that I visited the Test Valley in Hampshire for the first time. I was struck by the beauty of the river, its picturesque villages, cottages, farmhouses and small country houses with endless miles of meandering, white chalk walls with thatched or tiled copings. It was in this delightful part of rural England that I decided to make my home and to research the history of the local vernacular. This resulted in a dissertation prepared for the Architectural Association in 1982 following two years post-graduate study in building conservation.

My interest in chalk buildings led to visits to the New Forest to examine other buildings built of earth, but based on clay, sand and gravel. Geographically, the chalk and clay belts lie side by side but the style and method of building was different because of the individual nature of the two materials and the way the land had been utilized.

It soon became apparent that many of the earth buildings were in an appalling condition and that they were being demolished rather than repaired. Where attempts had been made to repair them, unsuitable modern materials had been applied by local tradesmen with good intentions but with little knowledge of traditional techniques. This often hastened the decay. It became quite plain that if the beauty of these two lovely areas of Hampshire were to be conserved, to delight my grandchildren in the way they had captivated me, then advice was needed urgently.

In the last decade I have visited, examined and reported on over one hundred earth buildings in need of repair, working mainly through the auspices of the Hampshire County Council acting as a grant giving

David Moore of Scolton Manor Museum, Haverfordwest
Gerallt Nash of the National Museum of Wales
John Owen of the Coliseum Museum, Aberystwyth
Clive Redshaw of North Kesteven District Council
Reg Reeves of Lincoln
Will Reynolds of Llanfallteg
Cormac Scally of the Ulster Folk and Transport Museum
Peter Skaife of North Kyme
David Smith of Leicester.

I would like to record my appreciation to other writers on the subject, all of whom are specialists in their own topic and whose valuable work has been published in books, magazines and society journals referred to in the bibliography. Many are known to me and my thanks go to Martin Andrew, John Ashurst, Nicola Ashurst, Dirk Bouwens, Dr Ronald Brunskill, Kevin Danaher, Jean Dethier, Ruth Eaton, Alexander Fenton, Jude James, John McCann, Martin Meade, M. Seaborne and Dr Bruce Walker.

Extracts from BS 5930 (1981) are reproduced with the permission of BSI. Complete copies of the Standard can be obtained from BSI Sales, Linford Wood, Milton Keynes, MK14 6LE.

Extracts from the CIBSE Guide Section A3 are reproduced by kind permission of the Chartered Institute of Building Services Engineers.

Finally, I would like to thank the conservation officers of the Department of the County Planning Officer and of Hampshire's District Councils for providing the live projects upon which so much of my experience has been gained.

Acknowledgements

It would not have been possible to write this book without the assistance of my colleagues within the office of Colin Stansfield Smith CBE, MA Dip Arch (Cantab) ARIBA, County Architect to the Hampshire County Council, and I am grateful for the interest, encouragement, assistance and support provided by the mechanical engineers, architects, structural engineers and librarians. In addition, I would like to record my thanks to the following individuals who have assisted me during my research.

Helen Albery of Sevenoaks
John Bostock of the National Trust
Dr R. Brunskill of Wilmslow
Ian Constantinides of Dorchester
Rodney Cousins of the Museum of Lincolnshire Life, Lincoln
Penny Copland-Griffiths of Trowbridge
Richard Dean of Salisbury
Michael Drury of Salisbury
Mr and Mrs R. Elborn of Sotby
John Fidler of English Heritage
Robin Freeman of Winchester
Ray Harrison of English Heritage
Peter Hood of Kettering
Alan Hounsham of Winchester
Alf Howard of Down St Mary
Larry Keefe of Teignbridge District Council
Eric Lewis of Wimborne

authority to owners of listed vernacular buildings and those in conservation areas. Examination soon revealed that traditional building materials were more suitable for repair work than modern materials and the increased demand for them has given rise to the establishment of a local supplier. It also became obvious that some building contractors were more sympathetic to the retention of vernacular buildings than others, and were keen to take advice on repair techniques and to experiment with traditional materials. A small but reliable workforce has therefore emerged in whose competent hands repair can be entrusted.

It is my hope that this book will encourage a sympathetic approach to the repair of many of our traditional buildings which play such an important part in the heritage of these islands. If the beauty of Britain's rural buildings is maintained to delight the next generation, then my task will have been worthwhile.

Gordon T. Pearson
May, 1992

Introduction

This book is the culmination of many years research, examination and experimentation. It is aimed at surveyors, architects, building contractors, building control officers, conservation officers and owners who feel that they are able to maintain their own properties with guidance. Within its pages I have attempted to answer every question that has ever been put to me regarding conservation, repair, alteration, conversion, extension and demolition. I have brought together much of what has been written on the subject, to which I have added repair and conservation techniques which I have noted in my sketch book as having been particularly successful. In addition, I have passed on experience gained over many years to present a comprehensive guide to earth building in the British Isles and its conservation and repair.

This is the first major book on this subject to be published in the United Kingdom since Clough Williams-Ellis and John and Elizabeth Eastwick-Field collaborated to produce *Building in Cob, Pisé and Stabilised Earth* in 1947. Reading that book once again, I was pleasantly surprised to find that much of the information is still current and that the views expressed still find favour today. It marks a milestone in the understanding of earth buildings in this country and is recommended further reading.

Earth has been used for many building purposes other than for load-bearing walls. It was used for floors, for wattle and daub panelling with wide local variation, and on roofs in Ireland both as an underthatch and as anchor points known as 'scollops' to secure thatching pins. This book restricts itself to the use of earth for loadbearing walls but has ventured into other areas where it is considered to be appropriate.

The approach I have adopted is designed to enable the reader to understand the material with which he is dealing, the different ways in which it has been used, its limitations and the importance that these points are understood fully before any attempt at repair is made. The reader is therefore urged to study the entire book and to acquaint himself with the way in which an earth building 'works' before adopting a cautionary approach to repair.

Furthermore, I have written this book from a practical, rather than scientific or academic, point of view, to enable the reader actually to construct or repair a building. It is my hope that other practitioners will adopt the advice given in this book, improving the techniques described and devising alternative solutions to common problems, thus ensuring the survival of Britain's most traditional of buildings.

Glossary

Italicized words also appear as main entries in this glossary.

Arris The sharp angle at the external corner of two adjacent wall surfaces.

Battered Describes a wall, the outer face of which is built with a backward slope.

Blinding See *sand blinding*.

Borrowed light A glazed window or screen which allows daylight to enter a room having no surface to the outside air.

Bund wall A low retaining wall often built against the base of an existing weak wall to increase its strength and to prevent movement.

Centre A curved temporary support, usually of timber, upon which an arch or vault is constructed. It is removed once the mortar has set or the *keystone* has been inserted to make the permanent structure self-supporting.

Collar A horizontal roof member tying together the rafters at their mid-points.

Collared roof A roof structure comprising *rafters* connected only by *collars* and *tie beams*. The load from a collared roof is distributed through the *wall plates* onto the front and back walls of a building.

Crucks Long curved timbers, framed together in pairs which rise from ground level to support the purlins of a building. They are usually joined together by a *tie beam* or *collar* to form an 'A' frame.

Distance piece A spacer in the form of a short length of metal tube threaded over a bolt to prevent a flush fixing.

Dormer window A vertical window constructed within the slope of a roof, either standing on the *wall plate* or on a trimmer joist framed into the roof structure.

Dovetailed fixing block A length of timber with tapered sides cast into a wall with the narrower face flush with the surface.

Flemish bond A method of bonding bricks characterized by alternate courses of stretchers (i.e. bricks laid with the longer face exposed) and headers (i.e. bricks laid with the shorter face exposed). The minimum thickness of a wall built in Flemish bond is therefore equal to the length of a brick (i.e. one brick thick). See *snapped headers*.

Footing The bottom part of an *underpin course* which has been built thicker than the wall to distribute its weight over a larger area of ground.

Formwork *Shuttering* of timber or metal used as a temporary support for the casting of concrete or for the ramming of earth.

Gable The vertical upper part of the flank walls of a building within the roof space, triangular in elevation.

Hip The sloping ends of the roof structure of a building. The hips rest on top of the flank walls at *wall plate* level. Where the hips rest on top of the flank walls above the level of the eaves wall plate, they are known as 'half-hips'.

Hydrated lime Non-hydraulic, high-calcium lime putty which has been dried and ground to a fine powder.

Hydraulic lime Limestone containing *pozzalanic additives* fired at temperatures in excess of $1000°$ C to produce a clinker which has been ground to a fine powder. When in contact with water, a chemical reaction gives it the ability to harden.

Igneous rock Rock formed by the cooling of molten lava from volcanic action.

Keystone A wedge-shaped stone or brick inserted at the top of an arch or vault, the final piece to be placed in position, signifying its completion.

Lift line A horizontal line visible on the face of an unrendered earth wall denoting the interval between the completion of one course and the commencement of the next.

Mattock A type of pick with a flat blade instead of a point, designed to be used with both hands. See *paring*.

Outshot (or outshut) A lean-to structure against the external wall of a building, usually sharing an extended roof slope.

Overhand A wall built from one side only, e.g. against an earth face.

Pantile A large, thick, roofing tile, one long edge of which overlaps the adjacent tile and whose other long edge is overlapped by the other adjacent tile. Sometimes slightly 'S'-shaped in section to facilitate overlapping and usually having a downward pointing rib at the top to prevent slipping.

Paring The action of trimming away surplus material from the face of a newly built earth wall using either a *mattock* or a purpose-made paring iron.

Perpend The vertical mortar joint between the faces of two adjacent bricks or blocks.

Pozzolanic additives Materials such as pulverized fuel ash (PFA) or finely ground high temperature insulation (HTI) fireclay powders, which produce an *hydraulic lime* when added to limestone. Traditionally, volcanic ash, known as 'pozzolana', has been used since Roman times.

Puddled earth A mixture of clay, sand and straw or chalk and straw, to which water has been added to turn it into a sticky dough.

Pugging A method of *wedging and pinning*, but using mortar lightly *tamped* into position to fill the space.

Punning See *tamping*.

Purlin A horizontal beam running along the length of a roof slope and supporting its *rafters*.

Purlinned roof A roof structure comprising *rafters* usually connected by *collars* and *tie beams* but supported on *purlins*. The load of a purlinned roof is transferred to the flank walls of a building. Where roof trusses are constructed to provide intermediate support to the *purlins*, the weight of the roof is distributed over all walls of the building.

Quoin The external corner of a building.

Rafter The sloping members forming a roof structure.

Relieving arch An arch built over a lintel in buildings of masonry construction to give additional support to the wall over an opening. Such an arch is described as 'rough' when built with rectangular bricks set in wedge-shaped mortar joints, and 'gauged' when built with wedge-shaped bricks set in mortar joints of a constant thickness.

Sand blinding A light dusting of dry sand thrown onto a decorated or rendered surface which is still wet, leaving a rough finish either to provide a key for further treatment or to give additional protection.

Scutch hammer A type of small pick with a replaceable serrated flat blade instead of a point, and designed to be used with one hand.

Sedimentary rock A rock formed in layers by the accumulation of sediment at the bottom of the sea.

Shoring Temporary external supports used to prop a potentially unstable building to prevent movement while repairs are executed.

Shuttering See *formwork*.

Smoke hood An inverted funnel erected over a hearth to channel the smoke to a chimney or roof aperture. Usually timber framed with sides of wattle and daub and supported on beams spanning a room.

Snapped header A brick used with the short face exposed and which has been cut in half to give the appearance of a wall one brick thick. See *Flemish bond*.

Spreader plate A length of steel or timber used under the end of a beam to distribute its weight over a larger area of wall.

Striking (or struck) The act of dismantling *formwork (shuttering)* once the concrete has set or the earth compacted.

Tamping The act of consolidating earth within a confined space by lightly ramming with any form of tool.

Through stone A stone used for bonding purposes, whose length is equal to the thickness of a wall and whose ends are therefore visible on both faces.

Tie beam A horizontal beam spanning a building, which ties together the feet of an opposed pair of rafters.

Tusk tenon A type of mortice and tenon joint found in carpentry work where the tenon protrudes beyond the mortice and is held in place by a small timber wedge let into the end of the tenon.

Underpin course The brick or stone base to an earth wall intended to separate it from the ground.

Wall plate A longitudinal, horizontal timber member built into a wall to support the ends of floor joints, or bedded on top of a wall to support the ends of rafters.

Wedging and pinning Filling in the space above new materials introduced into an opening and having the effect of transferring the load from the existing structure. The work is usually executed with several thicknesses of thin slates, gently tapped in from the side.

Chapter One

Introduction to earth walling: the materials

LOCAL TERMS

'Clay' and 'mud' are words used throughout the country to refer to all types of structures whose walls are constructed of earth. Dialectal words, however, remain in local usage, 'cob' perhaps being the best known but limited to the south-western parts of England although in Cornwall, the traditional variation 'clob' is still heard. In central southern England chalk buildings are normally referred to as 'chalk mud' but early references to 'swallows built cottages' indicate that a number of local terms once existed. The 'wychert' (white earth) of Buckinghamshire remains in local use and in the north of England, 'clay dabbin' or 'daubin' is the dialectal name used in the Solway Plain of Cumbria, as well as 'dung wall'. In south west Wales the term 'clom' is heard, or 'tai clom' in Welsh, whereas in central Wales 'tai pridd' is more usual and in north Wales the phrase becomes 'tai mwd'. In Ireland earth buildings are generally referred to as 'tempered clay', but in Scotland 'dab' is the usual expression, although a variety of local names have emerged for the different methods of construction which are found there. In East Anglia the earth block walling found there is known as 'clay lump' although throughout most of the world it is referred to as 'adobe'. Doubtless other local words were once in common usage but not all have been recorded.

Throughout this book the term 'earth' has been used to refer to any wall built of earth, whether it be clay-based or of chalk.

THE CLASSIFICATION OF SOILS

The British Standards Institute classifies soils for engineering purposes in BSS 5930(1981) as fine, coarse, and very coarse. Silts and clays are classified as fine whereas sands and gravels are classified as coarse. Very coarse soils comprise cobbles and boulders. Although cobbles are occasionally found in earth walling, their use is restricted and can normally be dismissed as being irrelevant in most parts of the country. The extract from the classification shown on the following page is relevant.

CLAY-BASED EARTH WALLS

To remain stable, an earth wall requires two distinct ingredients *viz* a cohesionless aggregate and a cohesive binder. It also requires water for mixing but most of this evaporates during the drying out process.

The aggregate forms the mass. It is the bulk filler that provides the density of the wall and varies depending upon its locality but usually comprises excavated sands and gravels, although quarry waste and road scrapings are also known to have been used. To ensure a dense mass which will compact well and form a permanent bond, the aggregate should be well graded from approximately 50 mm down, so that all interstices are filled. This will include medium and fine gravels, coarse, medium and fine sands and silts, the sand probably making up most of the bulk as it can be consolidated with the minimum of effort. The silt acts mainly as a gap filler, as it is high in absorption but low in adhesion.

The binder provides the medium which ensures that each piece of aggregate, once thoroughly coated, will stick together permanently. Clay is the obvious choice, but it is an unstable material, expanding when wet and shrinking when dry. Clays vary considerably from one part of the country to another and, even within a locality, in the amount of expansion when wetted. The most suitable types of earth building are those which expand least of all thereby ensuring a stable wall of permanence.

The clay content of an earth wall rarely exceeds 30 per cent of the total volume and 20 per cent is more usual. Its content is deliberately kept to a minimum to reduce shrinkage and cracking.

Occasionally, an earth is found to which straw and water can be added to provide the ideal mix for building. More usual, however, is a combination of local soils mixed on site to provide a strong wall which will remain stable yet permeable in addition to offering long term durability.

BSS 5930 (1981) Table 6: Description of soils

Basic classification	Basic soil type	Grade	Range of particle sizes (mm)	Selected notes on identification
Fine soils	Clays		Below 0.002	Smooth to the touch. Exhibits plasticity. Sticks to the fingers. Dries slowly. Shrinks upon drying.
	Silts	Fine	0.002 to 0.006	Coarse silt barely visible to the naked eye. Little plasticity. Slightly granular or silky to the touch.
		Medium	0.006 to 0.02	
		Coarse	0.02 to 0.06	
Coarse soils	Sands	Fine	0.06 to 0.2	May be well graded for size or uniform.
		Medium	0.2 to 0.6	
		Coarse	0.6 to 2	Large grains easily visible. Little or no cohesion.
	Gravels	Fine	2 to 6	May be well graded for size or uniform
		Medium	6 to 20	
		Coarse	20 to 60	Large enough for the shape of the particles to be described.
Very coarse soils	Cobbles	–	60 to 200	Only occasionally seen in earth walling but often used in underpin courses.
	Boulders	–	Over 200	Use restricted to underpin courses.

CHALK WALLS

Chalk was the last sedimentary rock to be laid down, and was formed between 70 and 100 million years ago. In its pure state it consists of a foraminifera aggregate set in a cementing matrix of coccoliths. Fora-minifera are a group of minute single celled shellfish, each one about the size of the head of a pin whose calcareous shell is perforated by pores. Coccoliths are lime secretions of algae and are related to the seaweeds. They are much smaller than foraminifera, a dot just visible to the naked eye, containing about 20 coccoliths. Their form varies depending upon the type of algae from which they have been secreted and they act as both a filler and binder.

The chalk of England varies in depth up to a maximum of 500 metres and it was formed over a period of 30 million years. There are three distinct bands known as the lower, middle and upper chalks, with a layer of hard Chalk Rock separating the middle and upper chalks and a layer of Melbourn Rock separating the middle and lower chalks. In certain areas, the upper chalk forms up to about 80 per cent of the total and is divided into two types known as belemnite and micraster chalks, the former overlies the latter. Belemnite chalk is found only in parts of Norfolk and the Isle of Wight and is not ideally suited for building. Where it has eroded leaving micraster chalk at the surface, as in Wessex, it provides the perfect aggregate and binder combined, provided that it is well graded for a range of particle sizes. Many buildings built of pure chalk are to be found in this area. It is particularly suited to the rammed method of construction.

WYCHERT

Wychert is found along the northern edge of the Chiltern Hills in a small area of Buckinghamshire between the towns of Aylesbury and Thame (Oxon) ten miles away. Buildings and boundary walling constructed of the material are confined to a narrow band no wider than six miles. To the east of Aylesbury, a few isolated buildings survive around Bierton but these are on the fringe of the wychert area. By far the greatest concentration may be seen in the small town of Haddenham where curving walls line both sides of winding footpaths, to create a microcosm of the local vernacular.

If left unrendered, wychert presents a pale, creamy-beige colour giving rise to its dialect name meaning white earth.

Like the chalk walls of Wessex, wychert is strong and was used as dug from the ground, a natural mixture of decayed limestone and very stiff clay. Only water and fibre needed to be added, usually in the form of chopped straw.

WATER FOR MIXING

The addition of water is necessary to mix the aggregate and binder together. It slurries the clay, thereby ensuring that the surface of each particle of aggregate, regardless of size, is completely coated. It also ensures that the resultant mixture will remain an homogeneous mass but the quantity used in the mixing process is critical to the construction of the wall. If too much water is added, the mix becomes uncontrollable, it lacks strength, is slow to dry and increases shrinkage. If insufficient is added, the binder will not break down sufficiently to coat the aggregate, thereby producing a wall which is badly bonded and of little strength. Most of the water added to the mix will evaporate during the drying out process, leaving the coated aggregate stuck together to form the wall.

ADDITIVES

Fibre

Additives of many types have been used, all of which provided additional stability apart from fibre.

Fibre is found in most earth walls. However, if the aggregate and binder are ideal, fibre is not always necessary. Walls found in parts of the East Midlands have been built without any fibre due to the fact that the iron-based clay binder is stable. Other walls without fibre have been found in parts of Devon. It usually takes the form of one year old straw, oat or barley in preference to wheat, but animal hair, heather, moss, chopped rushes, hay, coarse grass, twigs, furze, flax, sedge, twitch grass and cow parsley have also been found in different parts of the country.

The addition of fibre to the aggregate and binder provides several functions. J. R. Harrison has researched the subject and in his paper 'The Mud Wall in England at the Close of the Vernacular Era' he considers that its primary function is to distribute shrinkage cracks evenly throughout the wall. It also makes the task of mixing the ingredients easier by allowing them to be turned over and lifted. It helps to hold the earth together once it has dried out by acting as a light reinforcement and assists the drying process by ensuring that it dries evenly.

Stabilizers

Different types of stabilizers added to the basic ingredients affect the stability of a wall in different ways. The purpose is to increase the strength of the wall. However, they also increase resistance to water which, in turn, affects both shrinkage and the volume of the wall. Stabilizers react in one of two ways. They can block the capillary pores thereby preventing the flow of moisture, or they can provide a water repellent skin on the surface of the film of water surrounding each particle of earth.

One of the most commonly found stabilizers is cow dung. This reduces plasticity and is found in those areas where the aggregate content tends to be high, such as in the Solway Plain. In this respect, it acts as an additional binder thereby improving weather resistance. When used in chalk walling, a chemical reaction takes place which forms a gel and provides a weak stabilizing solution.

Another stabilizer which may be encountered is lime, added in the form of lime putty. When added to a clay wall its effect is limited but if added to a chalk wall, the chemical reaction is considerable and early strength is claimed, thereby allowing walls to be built at a more rapid pace.

Ashes are known to have been added to walls in Ireland but the effect of this can only be to supplement the fine sand content of the soil thereby adding to the bulk filler. No chemical reaction will have taken place and the stability of the wall will not have been affected.

Animal products such as milk, blood and buttermilk were sometimes added to earth walls but their use was not common. Casein is obtained by curdling milk with acid, and the resultant coagulant is particularly suited for use as a weak stabilizer when used with chalk.

Its use as a flooring material was common in chalk areas in mediaeval times, compacting well and drying to a high shine. It could easily be replaced or reconstituted as it was worn down by the feet of the cottage dweller. The addition of animal blood to an earth mix will improve its waterproofing qualities but records of its use are scarce. The use of milk has been examined by Richard Hughes who has carried out considerable research on the subject, and which is published in his paper 'Materials and Structural Behaviour of Soil Constructed Walls'. He considers that the milk creates a bacteria which grows fronds throughout the wall thereby aiding the binder to stick the aggregate together.

Cements of different formulae have been used at different times, particularly in the last part of the nineteenth century once its use became more common. The patent for Ordinary Portland Cement was taken out in 1824 but it was only one of several cements which were in experimental use at that time and it was several decades before it rose to its eminent position and much longer before all of its competitors ceased production.

As a stabilizer to earth walling, its use has been largely restricted to North America where it has been dubbed 'Terracrete'. However, experiments have been carried out in this country, initially at Amesbury made by Mr W. Jaggard in the 1920s which included a house built of rammed chalk and 5 per cent cement. The addition of such a small amount of cement increases the durability and strength of the wall considerably. If used in conjunction with clay, the general rule was to use more cement if the clay content is high, reaching a maximum of 7 per cent. However, a minimum of 4 per cent addition will achieve the additional strength and durability desired for all normal domestic applications.

Other experimental stabilizers used in the late nineteenth century included various bituminous solutions. The most common of these was obtained from asphalt which had been used as a cement in ancient times, but coal tar solutions were also used, particularly in the United States where it was known as 'Bitudobe'. Only a 5 per cent solution was necessary to ensure a rigid structure but the bitumens were difficult to use.

The addition of cements and bituminous solutions turns a wall into a rigid structure which behaves in a totally different manner to one built purely of earth. Its use in the United Kingdom has been rare and usually more by way of experiment. It has never found general acceptance and the chances of finding such a building are remote.

Chapter Two

Methods of construction

THE TRADITIONAL METHOD

Man has devised several ways of constructing his buildings in the British Isles, some materials being more suitable to certain methods than others. The basic traditional or piled method, however, was once found throughout much of the country and is the most commonly encountered. In the New Forest, it was known as 'throwing'.

Sometimes the materials were dug and used as found, sometimes they were excavated from several local sources brought together and mixed on site in heaps. Water was then added and trodden in to ensure a thoroughly damp, sticky dough. The earth was lifted on forks and slapped down on top of an underpin course constructed of stone or brick, compacted, protected and allowed to dry. Once dry, the entire process was repeated until the desired height had been achieved. The excess thickness was pared down on either side so that each was made flush with the underpin course. Although flues were sometimes constructed of earth, brick flues are common since it was more convenient for the bricklayer to complete his work ahead of the earth waller (or 'mud mason' in Scotland) leaving the flue and chimney stack standing proud. Once the roof was constructed, the building was allowed to dry out before being rendered externally to provide protection from the weather. The traditional method of construction is shown in Figure 1.

Detailed descriptions of the construction of earth walls exist from all parts of the country, dating back several centuries and it is clear that the piled method differed slightly from one area to another. The only

Figure 1 The traditional method. Working as a four man team, two men trample the mixture of wet earth and straw, one lifts it and one places it in position in a 500 mm high course. The lift lines between the courses are clearly defined and openings have been formed during construction rather than cut afterwards (HMSO photograph, 1922).

common denominator would have been the fact that all construction work took place in the summer months to ensure that each course dried as quickly as possible and that moisture laden earth was not damaged by frost.

The construction was usually carried out by farm labourers employed by the estate owners as part of their annual cycle of work which was dictated by the weather. However, it is known that in some of the larger villages in Wiltshire, the local builder would construct the shell of a cottage for '£100, more or less' at the turn of the century. In other parts of the country, it is recorded that it was usual for a young man to build a cottage for his bride, presumably with assistance from other members of both families. Other records from Cumbria speak of the entire community combining their efforts to provide a home for a newly married couple and being treated to a party of eating, drinking and dancing in return for their efforts. The same practice existed in parts of Scotland but the partygoers provided their own drink.

Preliminary work

It is clear that plans to construct a building were made well in advance. In those parts of the country where the subsoil was difficult to excavate, the top soil was cleared in autumn to allow the winter frosts to break it up, thus easing its removal in late spring. This practice was adopted in the chalk areas of Wessex and in those areas of Devon where local knowledge of the clay subsoils made it a desired practice. The pit so formed was sometimes filled with rubbish produced during the construction of the buildings, but where the ground was suitable it was used to form a cellar and the building was erected around the pit. Examples of this exist in Hampshire, in Kings Somborne and Winchester. In those parts of the country where water was scarce, the pit was lined with clay and used as a water storage tank, rainwater from the building drained into it. Examples are known on the Hampshire–Wiltshire border. Depressions often seen alongside boundary walls bear witness to the fact that they were constructed from the material found nearest at hand, particularly in chalk areas.

Mixing

Trampling and mixing of the earth involved both man and beast. In Dorset, women were expected to play their part and in Wiltshire the men wore heavy boots which were claimed to have been specially made for the purpose, or they strengthened their boots with iron soles. It is also known that horses and oxen were used to break down the lumps.

The addition of water has a significant effect on the constituents of the mix. Clay is an impervious material yet porous, expanding when moisture is added and shrinking when it is dried. When damp, capillary forces retain the moisture in the flat, plate shaped pores of clay making it impervious. When dry, the same forces cause the pores to stick together thereby reducing its volume by shrinkage. Sand reacts in a different way, allowing the moisture to flow through the angular grains with little resistance due to the comparatively large interstices and thereby causing only minor expansion.

The mixing varied with the constituent materials. Chalk is a stubborn material requiring time to allow the water to soak in but needing little mixing. Where the materials were brought together from different local sources, considerable mixing was necessary to ensure that each piece of aggregate, regardless of size, was thoroughly coated with clay. Early records speak of the pile being turned and trodden several times or simply soaked and straw mixed in. In order to perform its task adequately, the straw had to be pulled apart, scattered and added slowly as the other ingredients were being mixed to ensure that it was interspersed evenly throughout the entire wall. Failure to observe this procedure resulted in small nodules of fibre, twisted together, which became soft pockets vulnerable to erosion in the finished wall.

Mixing appears to have been achieved with a variety of tools. Forks of different types were usually employed as the suction created when lifting earth with a spade made the task almost impossible. Hay pitching forks are commonly referred to for lifting the material, although local blacksmiths created tridents with long handles specifically designed for the job in areas of the south and west as hay forks were not strong enough. An example of the Devon pattern, with a slightly bent end, can be seen in the museum at Tiverton, whereas the type developed further east in the chalk downlands of Wessex was flat. In Ireland, the use of graips is recorded as having been the mixing tool, a graip being a Scottish term for a three or four pronged fork used for lifting dung or digging potatoes.

Building the wall

At least two men were necessary to build a wall; the first, man 'A', standing on the ground and the second, man 'B', on the underpin course. At low level, man A lifted the earth and turned it over, slapping it down on top of the foundation. More was added until he felt that the maximum safe height had been achieved. This varied enormously from one part of the country to another depending upon the materials, the moisture content and local tradition. It was then consolidated, either by settlement of its own weight or by man B stepping onto it and trampling it, paying particular attention to compacting it well with his heels. Meanwhile, man A continued with his task, working all the way around the building to return to his starting point. Man B followed compacting all the earth, squeezing it out so that it overhung the underpin course on both sides.

Having completed the first course, it was covered with straw and weighted to allow it to dry without being damaged by rain. The drying period varied depending upon the materials, moisture content, weather conditions and the height to which it had been built. Earth containing chalk, such as wychert, dried fairly quickly; pure chalk probably dried quickest of all.

Once dry, the straw protection was removed and a second course added in exactly the same way as the first, and usually to about the same height. As the wall rose higher, the method changed. Man A lifted the earth by swinging it up to the working level where man B, now also carrying a fork or trident, was waiting to receive it. Having transferred it from one fork to another, man B, standing on the wall with his back to the direction of travel, both slapped it down in position and either battered or trampled it to consolidate it. If battered at an angle of about 45°, a pattern of diagonal lines was produced. Where each lift was laid at such an angle, consolidation was fairly minimal.

If insufficient time was allowed between courses for the previous course to dry out enough to take the weight of the next, the wall slumped and bulged at the bottom. The only remedy was to stop work and allow it to dry before proceeding. Once the surface had been pared down, a dip in the lift line was produced.

Forming openings

Records show that openings were sometimes formed as the walls were constructed. Although regional custom varied, evidence exists to show that timber fixing blocks were usually fixed to the sides of door and window frames and left projecting for casting into the wall as work proceeded. Additional blocks were cast into the top of the previous course for fixing window boards. Two timber lintels with large bearings were placed side by side over the top of the opening and often incorporated within the next lift to provide additional stability.

Some buildings were constructed as 'black boxes' without any openings, and pairs of lintels were built in where openings were to be cut once the walls were completed. A piece of timber was probably cast into the wall each side of the proposed opening so that two slots would appear once they had been pushed out at the completion of the work. These slots enabled a two handled cross-cut saw with removable handles to be inserted. With one workman either side, the wall was sawn down both jambs to produce an opening of the required height, and the centre earth panel was removed to form the opening. Personal observation during demolition works has revealed horizontal saw marks still clearly visible behind door and window frames.

Several problems arose if openings were cut after the walls had been built. Apart from constructing excess walling, entry to the building could only be achieved with ladders and the opportunity had been lost to build in fixing blocks for the joinery. However, where openings were cut, door and window frames were usually nailed to the underside of the lintel and stabilized by the external render and internal plaster.

Provision for fixings

During the construction process, provision was made for fixing joinery to the walls. In some cases, small timber fillets were cast into the walls at regular centres all around the building just above the underpin course. They were later used to provide a firm fixing for skirtings but can normally only be found in late-constructed earth buildings.

When boundary walls were constructed, provision was made for fixing wires for fan trained fruit as earth is not an easy material to which to fix. Fixings should be avoided wherever possible. The regional pattern varies but timber blocks, about 75 × 75 mm, were cast into the wall through its full thickness and sawn flush with both faces during the paring process. The exposed ends provided a secure fixing both for wires and, in some cases, to secure the coping. Alternatively, holes were sometimes drilled through the completed walls and threaded iron rods inserted with rectangular plates on either side held in slight compression by nuts, around which the wires were attached.

When stables, byres and other farm buildings intended for animals were erected, provision was made to ensure that the animals did not erode the internal face of the walls by abrasion. In Wessex this was achieved by casting long, horizontal baulks of timber, of dovetail section, into the wall as work proceeded, the narrower face finishing flush internally. This was then used both for fixing vertical protective timbers and for securing pegs on which to hang harness and other equipment (Figure 2).

Finishing and protection

The sides of the wall were pared down at different times according to the records. Some early observers commented that each course was pared down before the next one was added, thereby incorporating the debris in the next course, whereas other observers noted that the entire operation was carried out once the wall was completed. In some cases, the wall was almost completed before being pared, and the debris used to construct the final lift. The work was executed with axes, mattocks, byre forks, saws or spades, depending on when it was carried out but such was the scale of earth building in south and south-western England that local blacksmiths fabricated paring irons to increase the speed with which the work might be done. Such irons comprised a flat blade about the size of a spade but with a sharpened bottom edge, which was attached to a pole. To use the paring iron, the builder stood on top of the wall, held it vertically and chopped downwards, to trim off the excess earth. It appears to have been more suitable for use when each course was pared down before building the next course. Paring appears to have been carried out mainly by eye, and a plumb bob was used only to indicate approximate verticality. In most cases, the surfaces were pared back flush with the face of the underpin course to

Figure 2 The shell of a stable block damaged by fire showing details of construction. Dovetailed timbers were cast in to provide a combined harness rail and door lintel and to support the upper floor joists. Additional timbers provide a firm fixing for the door frame.

allow rainwater running down the face of the subsequently rendered surface to be shed clear. The overall effect of the pared surface was to expose the core of the wall. This accentuated both the lift lines and the differences in the proportions of the mixes used for each course, usually evident by different shades. Around openings which had been formed during the construction process, the paring operation exposed the vulnerability of the arrises, and it was easier to round them to prevent damage. The same problem arose at the quoins of the building, particularly in those regions where local sands were unable to take the wear. Once the paring had been completed, the wall was allowed to dry out to achieve maximum strength and an attractive, undulating surface of considerable character was exposed to view. This surface was not highly resistant to the weather and it was usual to protect it, particularly if it was a building as opposed to a boundary wall.

The protection usually took the form of a render but the wall had to be dry before it was applied. It often took at least a year and probably more for the wall to dry out completely and during the intervening period, it was exposed to erosion. Temporary protection in the form of straw matting hung from the eaves is claimed to have been used.

Allied construction work

Upper floor timbers were built in as work proceeded, the ends of which were sometimes tarred to prevent damage by damp walls. However, to assist in ensuring a level floor and to distribute the load more evenly, wall plates were sometimes cast in as the work proceeded. In some regions, the upper floor was built thinner than the ground floor so that the ends of the joists were not built into the wall but simply rested on wall plates which were stabilized by the floor boarding, thus ensuring that they would not be damaged by dampness in the wall.

Roof construction also varied from one locality to another depending on custom, the size of the building and the use to which it was to be put. Wall plates were usual, fixed flush with the out face of the wall, and the rafters nailed to them. Earth infill was added at the side of the plates and sometimes between the rafters, both to provide additional stabilization and to seal out draughts. The eaves detail varied and where wall plates were not used, ceiling joists were used to form large triangular trusses with each pair of rafters, the ends of which were buried in earth to provide stability. Gables are not common, and hipped or half-hipped ends are more usual to support the purlins. Where gables appear to exist externally, an internal inspection usually reveals a timber framework with brick infilling. Tradition varies, however, and lightly loadbearing gables are occasionally encountered.

Where dormer windows have been constructed, they form a building of one and a half storeys in height with a collared roof to provide adequate headroom. The simplest way to build them is to mount them on top of the wall plate, entirely within the roof space. However, in certain parts of the country, openings were formed or cut into the top of the walls and the dormer dropped into position, thereby breaking the continuity of the wall plate. The former method provides greater stability to the walls as any opening will weaken them.

Boundary walls were always finished with a coping to protect them against erosion by rain. Thatch is the traditional material to use but in many parts of the country it has been replaced by plain tiles, pantiles, slates or corrugated steel sheeting held down with iron straps fixed to timber blocks cast into the wall during erection.

Local variations

The construction of clay walls in Cumbria led to the development of a variation on the traditional method of building. Instead of building each course of clay to its maximum height and allowing it to dry for several days before adding the next course, thin courses were built with alternate layers of straw in a continuous form of construction.

The local drift deposits of sand and aggregate are of an angular nature and have a distinct red colour. They include cobbles as large as tennis balls which were occasionally incorporated in the work together with a binder of boulder clay and dung. For this reason, buildings of clay were sometimes referred to as 'dung walls'.

Once the walls had been pared down, the effect of the continuous method of construction could be seen more clearly and gave the appearance of many thin undulating layers of irregular courses of brickwork.

So rapid was this method of building that it became possible to build a cottage in one day with communal effort. Records show how the work was organized, with separate groups of volunteers digging the clay, wheeling it to the site, fetching water from the pond or ditch, mixing it and building the walls.

The development of constructing in thin layers probably arose when the builders realized that when a layer is compacted, only the top

Figure 3 Burgh by Sands (Cumbria). Large boulders were often used to construct underpin courses and to support the posts of cruck framed buildings. The Solway Plain once contained many clay dabbins which were built by the traditional method but to shallow courses to allow continuous operation. This complex of fine farm buildings is a rare survivor (photo: J. R. Harrison).

100 mm (approximately) is consolidated. It would therefore be better to build in thin layers thereby consolidating the entire wall rather than to build in thick layers and only consolidate the top part of each layer. Much of the consolidation is natural, being accomplished by its own weight as the water content slowly dries out.

Similar records of a continuous process exist in other parts of the country but it was not usual except in Cumbria (Figures 3 and 4 show

Figure 4 Detail of the shallow coursework of Cumbria. A layer of straw between each course of clay enabled the wall to dry out rapidly. Records exist of cottages having been built in a day using communal effort (photo: J. R. Harrison).

buildings in Cumbria). Isolated examples of continuous thin bed construction are known in County Limerick. In parts of Devon, where the subsoil was stable, lifts of more normal height were known to have been built with straw interleaving by a continuous process. Investigations into the wychert of Buckinghamshire have also revealed that the high strength of the material allowed it to be constructed without waiting for each course to dry.

In those parts of Wessex where pure chalk was used to construct walling, the lift lines were sometimes reinforced with hazel twigs 600 mm to 1000 mm long laid horizontally between the courses. They are not visible on the completed work and come to light only during demolition.

RAMMED EARTH

At the end of the eighteenth century, new methods of construction were introduced alongside the existing method and the pace of construction increased rapidly, fuelled by the introduction of the brick tax. Following a study of the French methods of earth construction, as recorded by François Cointeraux, Henry Holland translated the work into English and introduced the rammed earth or 'pisé de terre' method to the British Isles. This was heralded by the erection of experimental buildings for the Duke of Bedford at Woburn (Bedfordshire) by Holland in 1797.

An underpin course was constructed in the same way as for any other earth wall. Very strong, timber shuttering was then erected either side of the underpin course to a height of up to one metre and tied together with timber cross bearers or wires. Earth was thrown in and rammed in thin layers until the shuttering was filled. It was then struck, raised and re-erected where the process continued until the desired height had been reached.

This method of construction is totally different to the traditional method and no comparison should be made between the two. Rammed earth produces an artificial rock, well described by Clough Williams-Ellis and John and Elizabeth Eastwick-Field when they claimed that if 'particles of soil are forced together into close contact by ramming, they are compelled to adhere together by molecular and capillary forces and to form a hard and solid mass which is excellent for building purposes'. It normally contains no form of additive, not even fibre, and in most cases contains sufficient natural moisture to ensure adequate compaction.

Shuttering

It is sometimes thought that wattles might have been used for shuttering, but they are quite unable to take the tremendous pressure placed upon them during construction. Several strong, close boarded timbers were laid on edge horizontally abutting each other and they were held in place with vertical battens fixed on the outside at regular centres. Strong cross bearers were fixed at one metre centres both at the base and near the top which went right through the boarding to be held in place with wedges of tusk tenon design. Designs varied, some involving iron cross bearers and others with iron plates but all were of sturdy design able to take the strong forces acting on them. Additional supports were needed at the corners, and blockings were placed in the shuttering to allow door and window openings to be formed.

Filling and ramming

Damp subsoil was excavated, lumps were broken down into pieces and placed in heaps around the building. If the soil was too dry to adhere, a little water was added and allowed to soak in. Conversely, if the soil was too wet, it was allowed to dry until it was in the right condition. A well graded soil to ensure efficient compaction with sufficient clay to ensure stability was the aim. Chalk fulfilled these requirements without any further additive and became the most commonly used subsoil for rammed construction (see Figure 5).

Tools for the task were purposely made. Hardwood rammers weighed up to 9 kilograms each and had a vertical pole for a handle. Designs varied as experience had shown that a light rammer would not operate efficiently unless considerable effort was used and that a heavy rammer required too much effort. Iron rammers were introduced with heads of various shapes which determined the method of compaction.

Having oiled the inside of the shuttering to act as a releasing agent, a layer of soil was thrown in and levelled before being rammed by two operatives standing in the shuttering facing each other. Each man rammed alternately at an angle, starting at the outer edges and gradually working towards the centre before crossing their rammers to ensure complete compaction. Ramming ceased once the ramming tool no longer made an impression on the earth by which time the layer was about 100 mm thick. If the soil was too damp, it pugged and if it was too dry, it would not adhere. Work continued around the walling until returning to the starting point. The process was repeated as many times as necessary until the shuttering was filled. The top layer was finished with an indentation towards the centre to allow the next course to bond efficiently and to avoid a horizontal joint.

Figure 5 The rammed method. The foreman supervises two teams of four men. In each team two men work in relays to fetch and fill the strong shuttering with damp subsoil and two men work in unison to spread, level and ram it in 100 mm layers (HMSO phograph, 1922).

Should there be insufficient shuttering to enclose the building completely, each wall or part of a wall needed to be constructed as a separate operation before striking the shuttering and moving it along the wall horizontally to allow the work to continue. When this was necessary, bonding between adjacent sections was achieved by ramping the end of the filling to an angle of 45° against which the subsequent filling could be rammed.

The shuttering was struck immediately it was filled. The wedges securing the shuttering to the cross bearers were removed and the cross bearers were gently tapped through from one side, leaving mortices to be filled as a separate operation later. Where wires were used instead of cross bearers, they were severed and either pulled out or left in the wall. If iron cross bolts were used, they were withdrawn and the mortices filled later. The boarded shuttering could then be taken down.

To construct the second course, the shuttering was refixed at a higher level, overlapping the work already completed. The cross bearers were replaced and secured with wedges to allow the entire process to be repeated. The builders, however, worked in the opposite direction around the building to ensure that any lines of weakness which might have developed by one man being more efficient than the other were eliminated. Bonding between vertical ends of adjacent work was staggered around the building to avoid creating vertical lines of weakness.

It was not feasible to construct gable ends due to problems adapting the shuttering so hipped ends are usual. Where gables appear to have been constructed, they are, on further investigation, inevitably found to be built of rendered brickwork.

Internal walls were constructed after the external walls had been completed. Special care was necessary, particularly at the junction with the external wall where heavy ramming might have caused the outer wall to crack. Slots were usually formed on the inner face of the wall to allow the internal wall to bond and to avoid a straight vertical joint.

Flues were not normally constructed of rammed earth due to the problem of adapting the shuttering but where they are to be found, a clay drain pipe was usually used both as permanent shuttering and as a flue liner. Brick flues are common, often connected to the walls with long nails.

Forming openings

Openings were always formed during the construction process rather than cut afterwards. The shuttering was blocked off at door and window jambs and dovetailed shaped timber fixing blocks lightly nailed to it so that they would remain embedded in the wall once the shuttering was removed. Timber lintels were cast in as work proceeded. The opening formed was ready to receive a door or window frame which was nailed to the fixing blocks and to the underside of the lintel.

Alternatively, long hoop iron ties were fixed to the back of the door and window frames and left projecting. The frames were placed in position in the shuttering with vertical strips of boarding either side to block off its entire width. The hoop irons then became embedded in the wall as the shuttering was filled and compacted.

Stabilized earth

A variation on the normal rammed earth method, sometimes known as 'plating' may occasionally be encountered. Additional heaps of soil were prepared which were stabilized with a little lime or cement to improve its strength. As the shuttering was filled, the stabilized earth was placed against the sides and the unstabilized earth was placed in the centre. It was all rammed together to form one homogeneous mass. The effect was to produce a wall with a hard surface which did not dust when scratched. Above and below openings, stabilized soil was sometimes used for the full thickness of the wall to strengthen it at its bearing points and to provide ties across the openings. If the entire outer face of the wall had been stabilized in this way, it was necessary to stagger the thickness of alternate courses of stabilized earth to avoid a vertical straight joint between the stabilized and non-stabilized sections which might separate at a later date.

It is not known if this practice was common in the British Isles as there are no obvious signs of it in the finished building once it has been rendered. However, records exist of one such building having been erected in East Anglia where local architect B. G. Gaymer experimented with rammed earth early this century. At North Walsham (Norfolk), he built a house with walls of rammed loamy clay on top of a concrete underpin course and stabilized all vulnerable spots such as window reveals and the top of the wall at eaves level to improve stability. The long term durability of the building has not been recorded.

Location variations

Wessex

Rammed chalk buildings are found in the towns and larger villages of Wessex but difficult to recognize as their rendered exteriors hide their structural material which can be mistaken for brickwork. They were constructed by building contractors and built to a high standard. The employment of an experienced contractor was probably necessary due to the skills of carpenters being required to design and erect the strong shuttering necessary for this method of construction. Rammed chalk was used exclusively for housing those of a higher social class and fine residences were designed by local architects.

The walls were built vertically both sides, often with a set back at first floor level internally. Externally, they have a flat appearance with right angled corners often with mouldings worked into the render around door and window openings. The shallow pitched roofs are slate covered with hipped ends. Window openings were built larger than in traditional earth construction reflecting both the higher strength of the material and the fashion then prevalent in the towns in the nineteenth century. The window jambs were formed with an internal splay to allow as much sunshine as possible to enter the rooms. This feature is not generally found in buildings of the period constructed by the traditional method.

In Winchester many buildings were constructed in the early Victorian period when railway cuttings were being excavated and the problem of disposing of the surplus excavated material arose. Houses were often large and sometimes three storeys high built off a basement wall of random chalk rubble.

Such was the scale of building in Winchester at that time that developers started to experiment with different methods of construction which are not immediately apparent. Some buildings were erected by a method whereby fist sized lumps of chalk were picked out from the rest of the excavated material and put to one side whilst the rest of the material was broken down and used for rammed work. Shuttering was erected above the underpin course and the chalk lumps placed in position in a single horizontal row on either side, the space between being filled with puddled chalk which was gently tamped into position. The process was repeated, one row at a time, until the shuttering was filled. It was allowed to dry out sufficiently for the shuttering to be removed and re-erected at a higher level to enable work to continue. The finished appearance is that of random chalk nodules laid to courses and set in white chalk mortar. Once rendered, the buildings take on the appearance of a rammed chalk structure giving no hint of their constructional methods until alteration work is carried out.

Figure 6 Andover (Hampshire). A superb example of an early nineteenth century gentleman's residence built by the rammed method using chalk. The higher strength of the walls enabled larger openings to be constructed. The shallow pitch of the slate covered roofs is typical.

Figure 7 Winchester (Hampshire). A Regency villa constructed of rammed chalk showing the state to which the craft had developed. The tower is four storeys high but the timber lattice work is not structural. These houses were designed by architects and constructed by local building contractors. This fine example was demolished in 1983 to clear the site for a hospital extension (photo: K. Stubbs).

Another Winchester variation is the use of brickwork as permanent formwork to rammed chalk work. This tends to be used on the better quality buildings, usually if strategically sited. The brickwork would have been constructed well in advance to allow the mortar to set strong enough to allow the chalk to be rammed. Provision had to be made for fixing the shuttering, probably by wires built into the brick joints as work proceeded and held in place by plates. Once the shuttering was filled and rammed, it would be released by cutting the wires on the outside face of the brickwork and the wires pulled through on the inside. The shuttering was raised and the process repeated. The brick skin is usually in Flemish bond incorporating snapped headers and often built of yellow stock bricks.

Examples of rammed chalk work in Hampshire are shown in Figures 6 and 7.

A further variation to rammed chalk occasionally encountered is the construction of a permanent brick framework which was used both to support the shuttering and to strengthen the weaker parts of the building. Such buildings invariably have a brick underpin course and large brick columns, at least 450 mm wide on face, sited at each corner.

Further columns were erected to align with the jambs of the doors and windows. Once the mortar had set, strong shuttering, longer than the distance between the columns, was erected, using the thickness of the column as a template. Having filled the shuttering with chalk and rammed it, it was released, raised and refixed to enable the process to be repeated until the wall had reached the top of the columns. Occasionally, a horizontal band course at first floor joist level is encountered. The completed chalk panels were rendered and decorated, sometimes obliterating the brickwork. The village school at Grately (Hampshire) was constructed in this way but the entire building has been decorated thereby obliterating the effect of the chalk panels and contrasting brick framework.

Scotland

Records of 'clay and bool' buildings date back to 1650. They are also known as 'clay and dab' in Moray (Grampian) and 'Auchenhalrig work' in the town of that name.

A traditional mix of clay, mud, sand and gravel was prepared and mixed together with just sufficient water to provide a stiff dough. Fibre was added in the form of rushes or straw and small boulders or 'bools' were thrown in to allow them to become thoroughly coated. None were bigger than a man could lift with ease. In some cases, shuttering was erected on top of the underpin course and evenly sized bools were placed in rows against the outer faces. The space between was filled in with a type of dab but without the coarse aggregate bulk filler, and gently tamped into position to fill the shuttering and to ensure that there were no interstices. Every now and then, a bool was placed in the centre of the wall if space permitted, the deciding factor being the size of the bools in relation to the thickness of the wall which seldom exceeded 600 mm. Alternate layers of bools and clay were placed until the shuttering was filled. After allowing the wall to dry, the shuttering was struck, raised and re-erected to enable the work to continue, and a maximum of about 750 mm was possible in a day. Work continued until the desired height had been achieved when the final shuttering was struck and a blotchy wall of clay and bools was exposed to view.

Other accounts speak of the walls being constructed without the aid of shuttering, in lifts of 600 mm to 900 mm, each lift comprising several courses, with three or four days being allowed between each lift for the wall to dry before work continued.

Sometimes, the exposed face of the wall was rendered and decorated but if left to weather, the thin clay coating on the surface of the bools fell away revealing an attractive wall surface. The aesthetic effect was improved if each course of bools was set at a different angle to form a herringbone pattern. This was enhanced even further if the surface of the exposed clay was cut back to leave the bools projecting slightly and the joints pointed in white lime. In some instances, particularly in farm buildings, the outer surface was roughcast ('harled') with lime and sand.

Advantages of rammed earth

There are several advantages of the rammed earth method over the traditional method, the most important being the speed of erection. As the process is damp, as opposed to wet, it was continuous and reporters have commented that a team of three men could handle and place up to three cubic metres of soil in a single day. As there is no excess material to be removed by paring, time was not wasted constructing walling which was not required. The same advantage applied to the formation of openings. The higher strength of the rammed earth method meant that walls could be constructed more thinly and yet be immediately at maximum strength and able to bear the load. Walls built by the traditional method gain strength slowly over a period of a year or more before being able to take the maximum loading.

These advantages have an important effect on the cost of construction. The only known figures for comparison are contained in Walter Jaggard's report on his experiments at Amesbury (Wiltshire) in 1920 where cottages of different types of construction were erected and their costs compared with that of the 'control' cottage erected in cavity brickwork. The results were summarized in the table shown on the following page. According to these results, Jaggard found that it cost almost 25 per cent more to construct walling by the traditional method than by the rammed earth method.

Twentieth century experiments

Two of Jaggard's experiments are worthy of mention, *viz* Cottage no. 10 and Ratfyn 5. The former is constructed of chalk stabilized with 5 per cent cement and built off an underpin course of cavity brickwork filled with rammed chalk.

	Construction of wall	*Cost index*★
	Walls without finishings	
1	275 mm thick cavity wall comprising two half brick skins with 50 mm cavity, tied together with galvanized iron wall ties	10.00
2	375 mm thick solid wall of chalk built by the traditional method	6.19
3	375 mm thick solid wall of chalk and straw built by the rammed method	5.00
4	225 mm thick solid wall of concrete filled into shuttering	8.10
	Walls rendered externally	
5	250 mm thick cavity wall comprising two skins of stabilized chalk (5 per cent cement) blocks with 50 mm cavity, tied together with galvanised iron wall ties	8.10
6	375 mm thick solid wall of chalk built by the rammed method	5.48 to 6.19
7	175 mm thick solid concrete blocks	6.19
8	375 mm thick solid wall of stabilized chalk (5 per cent cement) built by the rammed method, each lift reinforced with strips of expanded metal lathing	8.10

★ Adjusted on a unit cube basis.

The walls are 425 mm thick on the ground floor and 350 mm thick on the first floor and incorporate rammed chalk flues. Ratfyn 5 is constructed of 70 per cent chalk and 30 per cent soil and built off an underpin course of brick and flint. The walls are of the same thickness as Cottage no. 10 and incorporate flues in the same way. Gable ends were constructed and it was necessary to adapt the shuttering to enable the work to be executed.

The main lessons learned from the experiments were that heart and wedge shaped rammers gave better results than flat ended rammers and that the best work was that which had been executed in the spring and the autumn. July and August were considered too hot and caused the chalk to dry out too quickly, preventing adhesion.

At the same time as Jaggard was conducting his experiments in Wiltshire, fellow architect Clough Williams-Ellis was carrying out a smaller scale exercise in Surrey. Using knowledge gained from research, he designed a bungalow of rammed earth at Newlands Corner near Guildford and supervised its construction. The external walls were constructed of rammed earth 450 mm thick and built off an underpin course of brickwork. Internal walls were of 225 mm thick precast earth blocks and the two were bonded together at junctions with long iron spikes driven into the rammed walls and left projecting for building into the block walls.

Following the success of Jaggard's and Clough Williams-Ellis's experiments, a number of other buildings were constructed shortly afterwards. At Princes Risborough (Buckinghamshire), a house of stabilized chalk was erected with the walls composed of cement and chalk in the very high proportion of 1:6. To provide even further stability, sodium silicate was added in lieu of water to provide a dense, solid wall which required no form of protection.

Gaymer also experimented with rammed earth at Paston (Norfolk) where he constructed a bungalow using his newly patented system of shuttering. The aim was to complete the construction of one wall of the building in one operation before commencing work on the adjacent walls. Piers of concrete blocks were erected at regular centres around the building, built off the underpin course. Boarded shuttering was erected on either side of the wall, kept apart by the piers and held in place by an independent framework which was connected at the top by long bolts. As the shuttering was filled, the boards were raised together with distance pieces spanning between them. Work continued until the top of the piers had been reached, the advantage being that the wall was not pierced by ties as would normally be the case in rammed earth construction. Openings were formed by temporary boxings placed in position as work proceeded.

Conclusion

Walls of rammed earth are stronger than those built by the traditional method. For this reason they could be built more thinly without affecting their loadbearing ability. They seldom exceed 450 mm thick on the ground storey and the upper floor is often built as thin as 300 mm or even less. Stabilized walls were built even thinner, internal walls sometimes being as thin as 150 mm.

Once completed, the walls were protected externally by means of a render and limewash in exactly the same way as if they had been built by the traditional method. However, the smooth surface formed by the shuttering was inadequate to ensure complete adhesion and it was necessary to hack a key on to the wall face.

Most buildings of rammed earth are found in central southern England and are built of chalk which forms an instant soft rock when rammed. In 1819, when the *Cyclopaedia or Universal Dictionary of Arts, Sciences and Literature* was published, Abraham Rees, describing the method of construction, commented that its use was, by then, common in the southern counties of England.

In these areas, rammed earth did not replace the traditional method of construction. The latter was usually the craft of the farm labourers in rural areas whereas the former was the trade of the building contractor in urban areas. The two forms of construction existed side by side.

ADOBE

At the same time as rammed earth was being introduced to the British Isles, the first adobe buildings were constructed both in Perthshire and South Cambridgeshire as has recently been discovered and reported by John McCann.

With the popularity of brick building in the Georgian period, it is perhaps not surprising that experimentation should take place to erect buildings of unbaked clay blocks, thereby saving both transport and fuel costs and enabling the unskilled worker or farm labourer to maintain the manufacturing process within his grasp, whilst working under the supervision of a semi–skilled man.

Adobe is a Spanish word meaning a sun dried earth brick, and soils which are suitable for other forms of earth construction are suitable for adobe construction. For this reason isolated examples of its use are known in Wessex, Wales, Scotland, the Midlands and in County Sligo and County Meath in Ireland, but by far the greatest concentration is found in East Anglia where boulder clays with chalk provide a suitable, stable medium and where it is known as 'clay lump' (see Figure 8).

The subsoil was excavated, placed in heaps and trampled to provide a well graded mix from about 25 mm down to dust in much the same way as in traditional earth construction. It was not usual to excavate each constituent material separately and mix them together. The excavated material usually contained a blend of all that was necessary to produce a good block but any deficiency could easily be remedied. East Anglian clay lump blocks tend to have a higher clay content than would normally be found in *in situ* walling. It was acceptable in block form as the shrinkage took place in the mould before the block was incorporated in the wall.

The natural constituents vary from sandy materials to almost pure chalk, both of which are suitable for adobe, but neither for brick making.

Both man and beast were used to trample the clay but as it is more economical to make a large quantity of blocks in one operation, the scale usually demanded the use of horses. Straw and water were added and mixed in thoroughly until a doughy consistency was reached. Occasionally, fibre in the form of sedge or twitch grass was added.

Figure 8 Adobe construction. Detail of a clay lump wall in East Anglia showing large, unfired earth blocks set in a lime mortar with thick joints. Part of the original mud render still remains which has been protected with coal tar.

The moulds were made of timber and varied according to local custom, small ones being capable of making only one block and larger ones capable of making several at a time. In either case, they comprised only sides, and maybe divisions, together with a handle at each end to allow removal. After wetting, the moulds were placed on the ground. The clay was thrown in, cut off flush with the top and lifted off ready for the process to be repeated. Block making was an efficient operation and it was claimed that two men could make up to 400 blocks in a single day.

There was no standard size for a block, they varied from 210 mm long × 100 mm wide × 60 mm high to 450 × 300 × 150 mm high. Sometimes, small vertical timber fillets were fixed to the inside of the mould which formed indentations on the finished block providing a key improving the adherance of wall plaster. The blocks were covered and left to dry for a few days before being turned to ensure even drying. This process was repeated, then the blocks were left for a few weeks until they became damp and ready for incorporation in the building work. During this period, considerable shrinkage took place.

An underpin course had already been constructed of flint or brickwork to at least 150 mm above ground level and ideally much higher. It was built only slightly less than the desired thickness of the wall to enable the subsequent rendering to finish flush.

The blocks were laid on top of the underpin course in stretcher bond, as in normal masonry construction, and bedded in clay or lime mortar about 20 mm thick. All walls were built solid, different thicknesses being achieved by laying the blocks in a different plane.

The rate of construction varied depending upon the size of the blocks being laid and the particular part of the building erected. However, it is known that at least one building contractor in Watton (Norfolk) based his calculations on a man and boy being able to lay at least 300 blocks, each measuring $450 \times 225 \times 100$ mm in one day.

External walls vary from between 225 mm and 300 mm thick depending upon the size and type of building, the former being most commonly used for housing work. Agricultural buildings such as barns have thicker walls, up to 325 mm being known, supporting a roof five metres high. Internal cross walls tend to be just under 200 mm thick, block bonded into the external walls.

The work continued as if in a brick type of construction, with upper floor joists built in, and openings formed with lintels, sometimes with rough relieving arches over them constructed in brickwork. Flues were often built of clay lump about 150 mm thick although brickwork was used for chimneys. Work stopped once the wall plate level was reached, and the last two courses were laid on edge to support the timber wall plate. Gable ends were commonly constructed as the blocks were easily cut along the raking edge.

Following construction and covering of the roof, the external surfaces of the walls were protected with tar or clay rendering and limewash. The interior was fitted out in the usual way, with skirtings and door linings nailed directly to the clay blocks.

Faced buildings

Buildings of clay lump construction have often been faced with brickwork. This was sometimes intended as part of the original design although some buildings have been faced as an afterthought, still retaining the external render. Even when constructed as part of the original design, the clay lump was usually completed before the brick skin was added, unlike cavity walling in which both skins are constructed together. Clough Williams-Ellis commented that he knew of occupied buildings awaiting the addition of the brick facing. Provision had to be made for the additional thickness when constructing the underpin course, whereas an added facing was built off an independent foundation and plinth. Provision also had to be made for tying back the brick facing to the clay lump. This was usually achieved with bent metal ties, one end of which was nailed to the clay, with the other end built into the brickwork as work proceeded.

Local variations

Buildings of chalk lump are also occasionally found. In 1932, at Quarley (Hampshire) newly qualified architect Jessica Albery constructed a house by this method which was named *Albery* but later changed to *Etekweni* (see Figure 9) before being destroyed by fire in

Figure 9 Quarley (Hampshire). *Etekweni*, designed and built by Jessica Albery in 1932 of chalk adobe. The blocks were made on site from crushed chalk and water and set in a mortar of chalk and sand. Miss Albery and Mr B. H. Nixon collaborated to construct several other similar buildings on the outskirts of the village during the following thirty years. *Etekweni* was destroyed by fire in 1987.

1987. Two years later, a pair of semi-detached cottages were built nearby, now converted into a single house. In later years, in conjunction with Mr B. H. Nixon, a local engineer, she constructed further dwellings in 1946 and 1960.

The underpin courses were constructed of concrete and some were topped with a damp proof course. The chalk blocks were made on site in moulds, laid on the ground. The moulds for the 450 mm and 300 mm thick blocks consisted of boards set on edge which were divided by bricks. The 100 mm thick blocks were cast in a small hand-worked moulding machine. The chalk slurry was made from chalk excavated on the site and mixed with straw and water in a pan mixer before being poured into the moulds. In the 1932 building, the walls were 450 mm thick constructed of blocks cast to a $450 \times 225 \times 225$ mm thick module whereas in the 1934 building, the external walls were 300 mm thick constructed of blocks cast to a $300 \times 150 \times 150$ mm thick module. The loadbearing internal walls were 225 mm thick constructed of $450 \times 225 \times 225$ mm thick blocks and non-loadbearing walls were 100 mm thick using $450 \times 225 \times 100$ mm thick blocks. The blocks dried quickly in good weather and could be incorporated in the works within 24 hours. However, the larger blocks tended to be brittle.

The blocks were laid in a mortar composed of ground chalk and sand although a little cement was added to the thinner partitions in the earliest houses. The flues were also built of the blocks but brickwork was used for the chimneys. They were lined with clay drain pipes and the fireplaces were lined with brickwork. The completed buildings were rendered using cement, lime and sand in the proportions 1:2:10 and decorated.

Adobe is not common in the chalk downlands of Wessex but a small estate of council houses are known to have been erected in Amesbury (Wiltshire) by a local contractor who stabilized the chalk by the addition of cement amounting to 10 per cent by volume.

A process for making building blocks of chalk was patented in 1937 by a Mr J. A. Cox who named it 'Durastone'. Production was intended to start in September of that year but its fate is uncertain so other chalk adobe buildings may await discovery.

Advantages of adobe

There are a number of advantages of constructing buildings in clay lump as opposed to the traditional or rammed methods. As the blocks are not incorporated in the works until they have dried out, all shrinkage takes place in the mould with the result that it is rare to find cracks in a clay lump building. As it is usual to make a large number of blocks well in advance of construction work, erection is rapid, and there is no waste and no further drying out. The manufacturing process is more easily controlled, under shelter if necessary, resulting in blocks which are stronger enabling thinner walls to be built using less material. The strength of the wall is increased further by the bonding of the blocks resulting in a durable structure once rendered and decorated.

PUDDLED EARTH

Puddled earth is a hybrid form of construction combining the materials of the traditional method with the shuttering of the rammed method. A traditional earth mix was prepared containing clay, silt, sand, aggregate and straw thoroughly mixed with water. Shuttering was erected on top of the underpin course and the mixture thrown in to a depth roughly equivalent to the thickness of the wall. It was lightly tamped into position to ensure consolidation. The drying process took several weeks after which the shuttering was struck, raised and re-erected. The process was then repeated until the desired height had been obtained.

The method should not be confused with rammed earth which used a damp soil without the addition of fibre. Rammed earth was a continuous process whereas puddled earth was an intermittent process and slower to construct than the traditional method. The rammed method involved heavy ramming whereas puddled earth needed only to be tamped into position, but the resultant wall was unable to match the strength of the rammed product. For this reason, a lighter type of shuttering was probably used.

Once completed, the building appears as if it has been constructed by the rammed method. The method of construction is often not apparent until part of the external rendering falls away to reveal the earth and its fibre. Sheltered, unrendered areas in outshots still reveal the grain of the timber boarding and where the wet mix has squeezed through badly fitting boards.

Puddled earth and rammed earth share the same advantages over the traditional method. However, the former can claim no particular advantage over the latter and its use in the British Isles has therefore been restricted. An example of puddled earth construction is shown in Figure 10.

Twentieth century experiments

Jaggard experimented with puddled chalk as part of his experiments at Amesbury. He erected a two storey house known as Cottage No. 4 Ratfyn. The walls were constructed of chalk and straw well mixed

Figure 10 Puddled earth construction. Two men fill the lightweight shuttering with a stiff mixture of subsoil, straw and water, whilst their two colleagues spread, level and lightly pun it to exclude air gaps. To ram such a wet mixture would cause it to pug (HMSO photograph 1922).

together with water. Shuttering was erected on top of the underpin course, and walls were constructed 425 mm thick on the ground floor and 350 mm thick on the upper floor. Both gables and flues were also constructed of puddled chalk.

Jaggard erected five buildings for his experiments. One was constructed of rammed chalk stabilized with cement, one of concrete, one of puddled chalk, one of rammed chalk and soil, and one of cavity brickwork. It is of interest to note that the consensus of those involved in the experiment considered that Cottage No. 4 Ratfyn was 'the soundest'.

Also in 1920, Robert O'Mennell experimented by building a chalet style bungalow for himself at Kenley (Surrey) in puddled chalk. The method of construction closely followed Cottage No. 4 Ratfyn at Amesbury, with the walls built of premixed pulverized chalk, straw and water thrown into shuttering and 'punned' in 450 mm high layers. The walls were 450 mm thick built off a flint underpin course and protected with render and limewash.

Buildings of puddled clay are known to have been erected in North Wales, the New Forest (Hampshire), all four provinces of Ireland and the southern part of East Anglia, but the method might have been more common than is generally known.

TIMBER DIAPHRAGM WALLING

Mud and stud buildings

Type and distribution

An interesting form of earth building is to be found in Lincolnshire, based on a timber framework which is completely covered with very thick daub giving the appearance of a building constructed by the traditional method.

Buildings can be found covering much of the county in an area from the Fens to the limestone belt but concentrated in the villages of the chalk wolds. About two hundred buildings survive between Boston in the south and Somercotes in the north, and between the east coast and Sleaford in the west. Isolated examples are known at Rempstone (Nottinghamshire), Thurcaston (Leicestershire) and Luddington (Northamptonshire).

Mud and stud buildings were once thought to be common in Leicestershire but most of those which survive often embrace several types of construction including open timber framework, cruck framework, mud and stud, wattle and daub and traditional earth construction. An example is known at Burton Overy.

The framework, usually of poor quality oak, comprised lightweight, roughly squared posts at irregular centres varying from 1.5 to 2 metres supported on stones forming part of a low underpin course. At the top they were mortised or housed into wall plates and pegged, and at the bottom they were connected by a sill beam on top of the underpin course. Strong, split staves were nailed to the outside of the framework to provide a rigid diaphragm for the daub. If the staves were slender, an intermediate rail was sunk flush into the face of the posts and pegged to give additional stability. The intermediate rails were staggered rather than in a straight, horizontal line and were provided only where the main posts were spaced well apart. The daub was applied to both sides of the staves in layers, finishing with a total thickness of up to 300 mm but usually less. The daub completely encased the staves and intermediate rails, just leaving the posts, wall plate and braces visible from the inside.

Various local names have been given to this method of construction including 'daub and stower' and 'post, pan and balks', the posts being the vertical timbers, the pan referring to the wall plate, sill plate and intermediate rail and the balks meaning the upper floor and roof timbers.

The traditional Lincolnshire cottage (Figure 11)

The method of construction lends itself to small, simple, single storey buildings and its most common use was for cottages although a number of small farm buildings such as shelter sheds and loose boxes were once known in the county.

The cottages have two or three rooms at ground floor level with a loft over and a central hearth with a timber framed, mud and stud clad smoke hood penetrating the ridge line centrally. In the earliest examples, the double pitched roof construction was simple and often consisted only of coupled rafters supported by the wall plates, held together by occasional collars. Ceiling joists spanned the width of the cottage, resting on the wall plates and supporting the loft floor. The flank walls were gabled and non-loadbearing. The later buildings incorporated purlins supporting the rafters and re-distributing the load over the four walls of the building. Half-hipped flank walls were introduced but many roofs appear to have been altered during the nineteenth century to improve the accommodation on the upper floor by replacing the half-hips with gables. A small window in the gable or half-hip gave a limited amount of light to the loft.

Mud and stud buildings often pass unnoticed because most were faced with brickwork in the nineteenth century. Only the elevations facing the most severe weather were protected in this way, and the limewashed daub is still usually visible on at least one elevation. At the

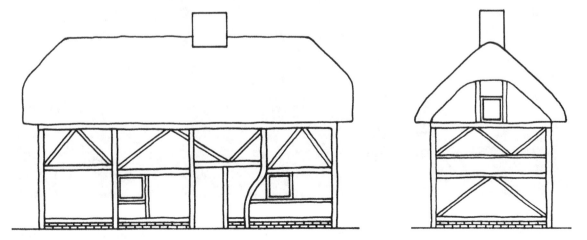

Figure 11 The timber framework of a traditional Lincolnshire mud and stud cottage.

same time, most timber framed smoke hoods were removed and rebuilt in brickwork.

During the last few years, several cottages have been repaired, some of which have public access. One example can be seen at Church Farm Museum at Skegness which was dismantled, moved to the site and re-erected in the early 1980s. The Old Vicarage Museum at Billinghay was built in the mid-seventeenth century but three of its walls were faced with brickwork in the nineteenth century. It was repaired between 1987 and 1989 by an employment training scheme organized by the local authority and is now open to the public. The original mud and straw floors are preserved under a later floor of mud, stabilized with lime, and are open to inspection. The framework is constructed of posts, wall plates, an intermediate rail and braces. The roof has half-hipped ends and has been repaired and re-thatched with an undercoat of reeds.

Figures 12 and 13 show examples with and without brick skins.

Recent arousal of interest in the repair of these buildings has given rise to publicity by the local media and examples such as Whitegates Cottage on the Gunby estate at Bratoft, which is owned by the National Trust, is open to visitors on a restricted basis.

Figure 12 Boston (Lincolnshire). Many mud and stud buildings were faced with a skin of brickwork in the eighteenth or nineteenth centuries making them difficult to recognize. The roof of Lincolnshire pantiles is traditional and the layout of the cottage is typical, including the high level window giving light to the loft.

Figure 13 Bagenderby (Lincolnshire). A fine example of a mud and stud cottage without a brick skin. The lightweight timber framework was clad externally with staves and daubed with clay to produce a wall up to 300 mm thick. The technique did not develop, making it difficult to date such buildings.

Mud and stud construction may appear to be a cheap type of building of only a semi-permanent nature but it was not considered to be so by local standards. It is known that buildings existed in the county in the sixteenth century in considerable numbers and examples still survive from the seventeenth century. Construction did not cease until well into the nineteenth century but as the method was never developed, the buildings are difficult to date.

'Stake and rice' and its variations

Several interesting methods of construction have developed in Scotland based on a timber diaphragm supporting earth cladding.

'Stake and rice' refers to one method whereby a structural skeleton of a building was erected of timber vertical posts (stakes), held together with horizontal twigs or oziers woven between them (rice, from 'hris', an old Norse word for twigs or brushwood). The framework was daubed with clay on one or both sides, and the timber merely provided a diaphragm for the daub.

Sometimes a rope of straw or heather was used in place of the twigs in which case it is more generally known as 'stake and tow' or 'stab and tow' in Angus (Tayside).

Another method known as 'caber and mott' is found just north of Inverness in Easter Ross (Highland). In this particular case, a framework of closely spaced vertical saplings was erected and the clay daub applied to both sides. In different localities throughout the area, this is also known as 'clay and mott', 'kebber and mott' or 'caber and daub'.

A similar system was sometimes used to construct partitions but the posts were placed further apart and horizontal members sprung into grooves at 600 mm centres providing a lattice framework ready to receive a clay daub.

REGIONAL DISTRIBUTION OF EARTH BUILDINGS

A glance at the map of the British Isles (Figure 14) showing the areas where earth buildings are to be found gives the impression that earth is a purely local building material. This was not the case. Had a similar map been prepared a hundred years ago, the areas of earth buildings would have encompassed entire regions, rather than isolated, detached areas which remain today. There is good and bad building soil just as there is good and bad building stone and surviving buildings are found only where the better soils exist.

Clay-based buildings are still found on a significant scale throughout most of Ireland, along the east coast of Scotland, on the Solway Plain, in the western half of Wales, the east midlands, East Anglia and in the south western peninsular. Chalk buildings are most common in Wessex but a few are found along the chalk belt, particularly in Norfolk. Wychert buildings are found only in Buckinghamshire close to the Oxfordshire border.

This does not dismiss all other parts of the country. Isolated earth buildings are still being located widely throughout much of the country providing historians with tangible proof of what the written records claim. Remnants exist in the Lleyn peninsular (North Wales), in Warwickshire, in the moors and wolds of Yorkshire and in the Suffolk, Hertfordshire and Essex areas.

There is little doubt that wherever suitable materials are to be found, earth buildings once existed. That they do not remain today is probably due to the lack of durability of the binder rather than the lack of construction skills. Earth buildings are therefore still to be found in

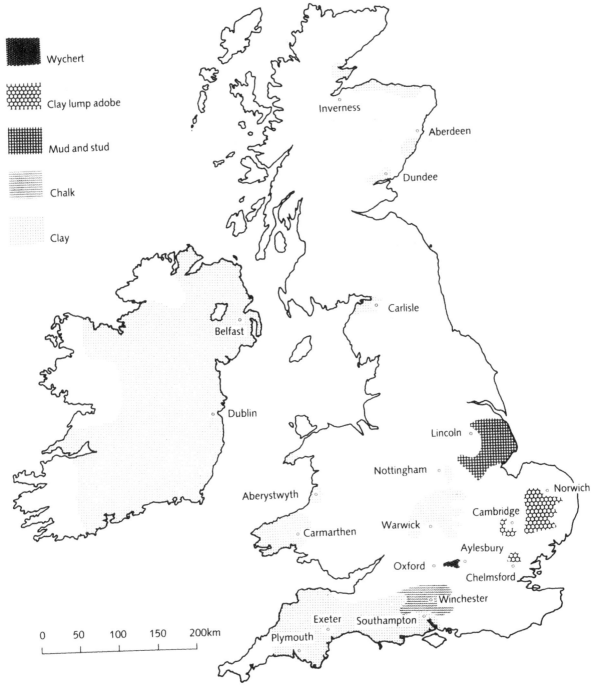

Legend:

- Wychert
- Clay lump adobe
- Mud and stud
- Chalk
- Clay

Inverness
Aberdeen
Dundee
Carlisle
Belfast
Dublin
Lincoln
Nottingham
Norwich
Aberystwyth
Cambridge
Warwick
Carmarthen
Aylesbury
Oxford
Chelmsford
Winchester
Exeter Southampton
Plymouth

0   50   100   150   200km

Figure 14 The regional distribution of earth buildings.

those areas where the natural materials are most suitable. The areas defy exact geographical boundaries, but they give a general indication where earth buildings can still be found in appreciable numbers. Isolated surviving buildings exist elsewhere and are still being discovered.

Access to buildings open to the public

Enquiries regarding access to those buildings open to the public should be made to the following organizations.

1. Skegness Church Farm Museum:
    Museum of Lincolnshire Life, Burton Road, Lincoln, LN1 3LY. Telephone 0522 528448.
2. Billinghay Old Vicarage Museum:
    North Kesteven District Council, Planning Department, 81 Eastgate, Sleaford, Lincolnshire, NG34 7EF. Telephone 0529 414155.
3. Whitegates Cottage (National Trust) Bratoft:
    By written appointment only: April to September.
    John Zaremba, Whitegates Cottage, Mill Lane, Gunby Hall Estate, Bratoft, Near Spilsby, Lincolnshire.

Chapter Three

The qualities
of earth walling

INTRODUCTION

The qualities of earth walling are varied and the purpose of this chapter is to provide guidance for their conservation, pointing out those factors which determine each quality. The material ranges from clay to chalk, both of which vary from one region to another, and, in some parts of the country, they are mixed together. Methods of construction vary from the traditional method to the rammed method, each producing different qualities in the completed walls; adobe is so different from either that it may be considered as a separate category.

Little research appears to have been undertaken on earth walling because it is not in regular use in the British Isles today. If it were to achieve a renaissance, facts and figures would be required by the local authorities building control departments, and the soils testing laboratories would respond to the demand for facts and figures regarding strengths, permeability, thermal insulation, and so on. Where statistics are available they have been given, although some are dated and have been converted to metric to allow comparison with more recently obtained information. They must be regarded only as a general guide, and flexibility should be used in their interpretation. The range of statistics quoted is wide.

Chalk, however, is the exception to the rule and much detailed research has been undertaken and published. Although chalk is not a major material for walling except in central southern England, it is used widely in earthworks for civil engineering projects throughout the chalk belt, and engineers require accurate information to design safely. Statistics for remoulded mass chalk have been quoted due to its similarity with chalk walling, particularly in rammed construction, although the exact nature of the similarity has not been established, due mainly to different degrees of compaction.

Information has been collated and presented under appropriate headings but there are no clear divisions between some of them and they may need to be considered together to be understood properly. For example, moisture content, plasticity, porosity and permeability are all closely allied. So are strength, density and durability. Even the insulating qualities of the material cannot be taken in isolation as this is dependent upon its density and moisture content.

MOISTURE CONTENT

The traditional method

Moisture content or water content is that mass of water which is capable of being extracted from the soil when a sample is heated and dried. It is usually represented as a percentage of the mass measured when dry.

It is the one vital statistic that ensures the stability of the wall during construction. If there is insufficient moisture, the earth will not mix or compact properly. If there is a little too much moisture, the builders will find their task easier but this will be reflected once the wall has been completed and shrinkage cracks appear which cannot be absorbed by the fibre. If the moisture is far in excess of what is required, the earth will not only be difficult to lift and place in position but will be incapable of standing on its own. It could also destabilize the previous course, causing the entire wall to sway and become unstable. This would necessitate complete rebuilding, although it may sometimes be corrected.

The fibre is capable of absorbing moisture and the continuous method of construction used mainly in the Solway Plain area with its shallow courses and straw interleaving layers might well indicate that these walls were built with an excessively wet mix. Trial and error taught the mud mason when he had added sufficient moisture, allowing him to develop rule of thumb guides which would have been passed on but not, unfortunately, recorded.

As earth walling is composed of a blend of aggregates and binders, the natural moisture content of each tends to differ. The moisture content of cohesive silts and clays is much higher than that of granular sands and gravels. To build a strong wall, a high density is required, but this can only be achieved if the moisture content is maintained at the minimum necessary to ensure that the clay binder coats each grain of sand and aggregate thoroughly.

Rammed and puddled earth

Walls of rammed clay were built with a low moisture content or the material pugged when compacted. One need only squeeze a handful of barely damp soil to see how easily it adheres together and to realize that recently excavated soil needs no additional moisture. However, if soil had been allowed to stand in warm, drying winds, a little water was probably added before it was capable of compaction. An optimum moisture content of about 10 per cent was usual if the earth was sandy and had a low clay content. This rose to 20 per cent if the clay content was high.

The moisture content of mass uncut upper chalk measured above the water table level can vary between 20 and 28 per cent, with the micraster chalk having a slightly lower level than the belemnite chalk. However, once the chalk has been excavated and reconstituted, the readings tend to drop by about 4 per cent. The ability of chalk to absorb large amounts of water is well known: the downlands of southern England are often referred to as a giant sponge. However, it absorbs water at a lower rate than clay-based soils. As with rammed clay, an excessive moisture content will render the chalk incapable of compaction and its optimum is probably between 15 and 20 per cent, decreasing significantly with increased compaction. This is well below the natural moisture content of the material but once the reduction due to excavation and reconstitution has been taken into account, it is usually at its optimum for efficient compaction.

When Jaggard carried out his experiments, he commented that Cottage No. 10 built of rammed chalk and stabilized with 5 per cent cement needed no further water as the natural moisture content of the chalk was 20 per cent. This produced walling which he described as 'thoroughly sound and workmanlike'.

When working with puddled earth, Jaggard prepared the chalk and straw as though he was about to build a wall by the traditional method, and added water amounting to 8 per cent of the total weight of the chalk. Knowing that the chalk already contained 20 per cent moisture, the total moisture content of the mix was 28 per cent of the weight of the chalk. When summing up his experiments he commented that the mix for this particular building, (known as Cottage No. 4, Ratfyn) gave the best results.

Drying out

Once completed, all earth walls adjust their moisture content to harmonize with their environment and it is during this period that shrinkage occurs and cracks develop. Earth walls built by the traditional method, whether of clay or chalk need a high moisture content to enable them to be built so their moisture loss is the greatest. The fibre copes with the shrinkage by spreading it evenly throughout many hairline fractures which can barely be seen. However, if the amount of moisture to be lost is too great, the fibre is unable to cope and a major structural crack will occur. This risk is greater with clay-based walls rather than with chalk but is dependent upon the expansiveness of the clay.

Walls of rammed earth react in a different way. As their moisture content was low to begin with, there was little to dry out so the risk of cracking was no less great. For this reason, the addition of fibre was not necessary.

Puddled earth dries slowly. Its moisture content was as high as walling built by the traditional method but the rate of drying was controlled by the insulation of the shuttering rather than by the surface tension generated by warm winds. The inclusion of fibre together with this slow drying out process produces a strong wall, second only to rammed earth. Both Jaggard and Mennell expressed their satisfaction with the walls they constructed in this way.

The length of the drying out process was dependent upon the moisture content, the thickness of the wall, its location, its degree of compaction, its fibrous content and the nature of the material. It could take up to two years before achieving an equilibrium of between 5 and 10 per cent. This figure varies from the surface to the centre of the wall and from one season to another. Although it varies little between chalk and clay, it is dependent upon the location of the building, both on a local and regional basis. The latter point is illustrated by comparing the map of the British Isles showing where earth walls are still to be found with the rainfall map. It will be noted that no earth buildings are located in areas where rainfall is in excess of 120 cms per year.

PLASTICITY

The Atterburg limits

The plasticity of a soil is dependent upon its type and its moisture content and is defined by the Atterburg limits, named after the Swedish soils scientist who suggested them. These limits can be represented in graphic form, as shown in Figure 15.

When water is added to earth, there comes a point when the volume of the earth starts to expand, this point is known as the *shrinkage limit*. Conversely, it can be thought of as the moisture content measured at a point when the soil ceases to shrink while being dried.

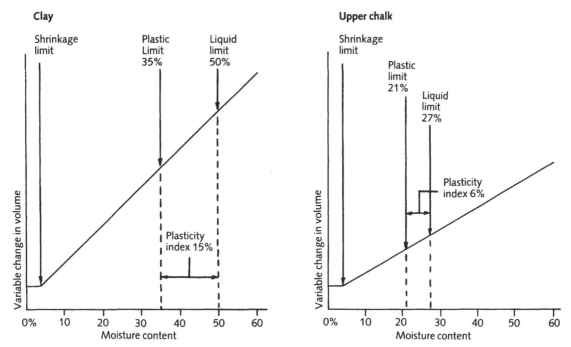

Figure 15 Comparison between typical samples of clay and upper chalk.

As the moisture content is increased, it is retained as a halo around each particle of soil until the soil is transformed from the semi-solid state to the plastic state, a point known as the *plastic limit*. Conversely, it is the moisture content measured at a point when the soil becomes too dry to be in a plastic condition.

If the moisture content is increased even further, it will reach a point where it leaves the plastic state and becomes a liquid, a point known as the *liquid limit*. Conversely, it is the moisture content measured at a point where the soil passes from a liquid state to a plastic state.

The British Standards Institution has set out guidelines to determine these limits and they are contained in BS 1377 (1990). The plastic limit and the liquid limit can be measured in terms of percentage moisture content, the difference between the two representing the range within which the earth remains plastic. This difference is known as the plasticity index.

Clay and chalk compared

Chalk and clay react in different ways when water is added. Clay particles are plate shaped and chalk particles are spherical and they each absorb different quantities of moisture in a different way.

Upper chalk is considered to be a non-plastic material but once it has been broken down and reconstituted, the coccolith matrix tends to dictate its behaviour. It is a single mineral with a constant positive electrical charge over the surface of its particles. Below the plastic limit, the moisture is retained by negative bonding thereby creating a stable state. Above the plastic limit, the excess water merely binds together the outer edges of the haloes rather weakly, abruptly turning it into a temporary liquid. This plastic condition is known as 'thixotropy'.

Clay possesses both positive and negative electrical charges over the surface of its particles and the plastic condition is established below the plastic limit, not above it as with chalk. The clay will continue to absorb moisture until the plastic limit has been reached, without the abrupt change associated with chalk.

The low plasticity index of chalk stresses the need to be vigilant when adding water for fear of exceeding the liquid limit, whereas the high plasticity index of clay allows more tolerance in the use of water before reaching its liquid limit.

Clay requires more water than chalk to reach its plastic limit, the former achieving it with a moisture content of 35 per cent and the latter at 21 per cent. Similarly, clay reaches its liquid limit at 50 per cent and chalk at 27 per cent. The resultant plasticity indices of 15 per cent for clay and 6 per cent for chalk illustrates the difference in the nature of the two soils.

The addition of fibre or stabilizers affects the figures considerably. It is thought that plasticity is reduced if fibre is added, depending upon quantity and type. A similar reaction takes place if dung is added to the earth.

The relationship between the material and its moisture content should be understood clearly. In order to achieve a plastic mix suitable for building, it is apparent that although clay requires more water to be added than chalk, this is more than offset by the low percentage of clay binder found in a typical wall of blended soils. With a clay-based mix, only the clay and silts will absorb moisture yet these amount to only about 25 to 30 per cent of the total mass. With chalk the entire mass will absorb moisture.

When a blend of clay and chalk is used, as in Buckinghamshire and parts of Wessex, the entire mass will absorb the moisture. However, the clay binder has the ability to absorb more than the chalk and may well do so thereby allowing the chalk particles to be coated much more thoroughly with clay than if sands and gravels are used. The fusion between the clay and the smallest of the chalk particles is complete and impossible to separate with simple soil testing techniques and could well be the reason for the high strength and durability of wychert.

POROSITY AND PERMEABILITY

Definitions

All soils allow moisture to pass but at different rates dependent upon their porosity. Porosity is the volume of voids (air and water) expressed as a percentage of the total volume of its mass. Permeability, also known as hydraulic conductivity, is the ability of a material to allow the passage of a fluid and is expressed in millimetres per second (mm/s). The two are obviously closely linked.

Moisture cannot pass through the particles, only between them so that soils with large voids will be more permeable than those with small ones. A well graded soil will ensure that all interstices are filled and the rate of permeability will be reduced to a minimum.

Factors affecting the permeability of clay soils

A wide variation should be expected when quoting rates of permeability. As a general rule, a clay wall with a low percentage of sand and gravel will have a low rate of permeability whereas a clay wall with a high sand and gravel content will have a high rate.

The shape of the clay particle encourages capillary action so that walls with a high clay content tend to be affected by rising damp to a greater extent than those where the clay content is lower. This can sometimes reach over 2 metres and put the ends of the upper floor joists at risk from dry and wet rot.

The height to which moisture will rise varies from one season to another and is affected by temperature. Warm air will absorb more moisture than cool air resulting in constantly changing rates of humidity. Cool air with a low moisture content will restrict the height to which damp will rise whereas warm, moist air will encourage the damp to rise higher.

There is a relationship between density and permeability. A well compacted wall producing small pores will restrict the flow of moisture and a poorly compacted wall producing large pores will allow a higher rate of flow. For this reason, walls of rammed earth are less permeable than those built by the traditional method.

Permeability is also affected by the direction of travel. The vertical flow is affected by capillary action and the horizontal flow by particle shape and external forces such as wind pressure.

These factors affect all forms of earth construction but an additional factor should be borne in mind when dealing with adobe. It is the only form of earth walling which involves jointing, and moisture will travel through the joints at a different rate to which it will travel through the blocks. A clay mortar is less permeable than the blocks but if a sand–based mortar is used, although the rate of permeability will normally be higher, it will be reduced by the stabilizing effect of the lime or any other additives.

Despite the many factors which affect the permeability of an earth wall, rates of flow varying from 10^{-1} mm/s for sandy soils to 10^{-5} mm/s for heavy clays have been measured and recorded.

Upper Chalk

Intact Upper Chalk rock is highly porous but its percentage porosity varies widely from one part of the country to another. In Ireland, porosity of less than 5 per cent has been recorded whereas in south east England it varies from 30 to 55 per cent. However, most chalks come within the range 40 to 45 per cent.

Despite its high porosity, its rate of permeability is low, ranging from 10^{-5} to 10^{-8} mm/s. This is probably due to the microporous nature of the material and the way the chalk retains its moisture within the foraminifera. Once excavated and reconstituted by the traditional method, the same density could not be achieved resulting in larger pores and greater permeability. However, when rammed into shuttering, a density approaching that of intact rock should be attainable with similar porosity and permeability qualities. Stabilization with cement or lime would reduce the permeability factors even further.

Condensation

Condensation is formed when warm air containing water vapour comes into contact with a surface whose temperature is lower than the air. If the colder surface is impermeable, for example glass, the water vapour will condense on the surface and run down to form a puddle. However, if the colder surface is permeable such as an earth wall, the condensation may form on the surface or at any point (the 'dew point') within the thickness of the wall where the temperature becomes lower than the air.

In the past, condensation was not a problem as buildings were heated by open fires which required a supply of air to operate efficiently. Ventilation from gaps around doors and windows was adequate to fuel the fire and to reduce the moisture content of the air. Radiant heat from the fire allowed the walls and flue to absorb warmth and to release it very slowly into the building when the outside temperature fell at night. However, the comfort of modern living demands the installation of a gas or oil fired central heating system which is capable of instant response. This encourages intermittent use in an attempt to save fuel, preventing the fabric of the building from being warmed. The risk of condensation is therefore increased. A low, gentle, constant heat which allows the fabric to maintain warmth is therefore preferred.

The two main sources of water vapour are the occupants of the building and the way the occupants use the building. Occupants perspire and breathe which produces moisture, and buildings may contain gas or paraffin oil fired stoves without flues, tumble driers, washing machines, shower units, etc. all of which produce large quantities of moisture. The warmer the air, the more moisture it can absorb and the greater the risk of condensation. This is reduced by ventilation but is adversely affected if doors and windows have been draughtproofed.

The importance of ensuring that nothing hinders the passage of moisture through the external walls is discussed in Chapters 6 and 7. However, provided that cement-based wall plasters and impermeable paints are not introduced, condensation should not be a problem in an earth building and calculations based on thermal conductivity factors are not therefore necessary.

THERMAL INSULATION

The preference for earth

Cottage dwellers throughout the British Isles have commented on the warmth of earth buildings and these have been recorded by early travellers. In the west country, S. Baring-Gould writing in 1923 commented 'I have known labourers bitterly bewail their fate in being transferred from an old fifteenth or sixteenth century cob cottage into a newly-built stone edifice of the most approved style, as they said that it was like going out of warm life into a cold grave'. Jonathon Binns (reproduced by Alan Gailey in *Rural Houses of Northern Ireland*) writing in 1835 recalled a visit he made to County Armagh in Ulster. He commented 'We called on an old man who was having a stone cottage built for him at Lord Gorford's expense. He said "I would rather live and die in my old cabin, the mud walls are warm, it is the warmest in all Ireland"'. A similar comment is recorded by P. W. Barnett (reproduced in *Building in Cob and Pisé de Terre* (HMSO)) when he visited the tenants of clay lump cottages on Sir Hugh Beevor's estate near Harling (Norfolk) which were constructed in the 1880s. He wrote 'The inhabitants of these dwellings say that, in spite of the fact that the walls are comparatively thin, their houses are cool in summer and warm in winter'.

Thermal conductivity and resistivity

The comments of the cottagers can be checked against heat loss calculation made in accordance with the CIBSE (The Chartered Institution of Building Services Engineers) Guide and compared with the minimum required by the Building Regulations. The latter have been revised in recent years reflecting both the rise in fuel costs and the need to conserve all forms of energy.

Thermal conductivity (or heat loss) through a material is represented by the Greek letter lambda (or more correctly, $\lambda\alpha\beta\delta\alpha$) in the form λ in accordance with the ISO Standard 31 Part 4 (1978) and this has been adopted by the British Standards Institution in BSS 5775 Part 4 (1979). It is represented by factors which are expressed as the amount of heat, measured in watts, which will travel through one square metre of fabric and integral air pockets one metre thick, when a difference of one degree Kelvin (centigrade) is applied to each side of that fabric.

Thermal resistivity is the reciprocal of thermal conductivity and is represented by the letter r.

The CIBSE Guide lists factors for all known building materials and those factors appropriate to earth walling are extracted and shown on the following page. It should be borne in mind that the lower the thermal conductivity factor, the better its insulating quality.

Both the thermal conductivity and thermal resistivity factors can be used to calculate heat losses (i.e. U values) through different types of earth walling taking into account surface resistances either side. The results are tabulated (on page 54) and compared with U values for other types of wall construction. A certain amount of interpolation has been necessary to obtain the figures for clay-based earth walls depending upon the amount of binder but the results vary so little that the degree of error can be considered small.

The Building Regulations which apply throughout the whole of the United Kingdom have recently been amended to improve the insulation of the walls of domestic dwellings. A U value of 0.45 W/m^2K is the minimum requirement for all new dwellings erected since the 1990 Amendment came into force. The Regulations were previously amended in 1985 when a U value of 0.60 W/m^2K was required. Reference to the tables shows that none of the types of walling listed comply with the latest requirements without further insulation.

The tables confirm the beliefs of the cottagers that earth buildings were warmer than stone. Stone is revealed as a poor insulator with little difference between different types of sedimentary rocks (limestone and sandstone). Igneous rocks (granite) are the worst insulators of all. Chalk and clay walls vary little and the varying percentage of

Extract from Table A3.22

Material	Condition	Thermal properties λ $\left\{\dfrac{W}{mK}\right\}$	r $\left\{\dfrac{mK}{W}\right\}$
Crushed Brighton Chalk	Dry	0.25	4.00
	10% dry weight	0.50	2.00
	20% dry weight	0.80	1.25
Mud	5% dry weight	0.43	2.30
	10% dry weight	0.72	1.40
	20% dry weight	1.44	0.69
	40% dry weight	1.15	0.87
	80% dry weight	0.94	1.10
	150% dry weight	0.79	1.30
Soil			
– clay 1500 deep	11% dry weight (i.e. typical)	1.10	0.91
– clay	11% dry weight (i.e. typical)	1.50	0.67
– clay, loosely packed	14% dry weight (i.e. typical)	0.38	2.60
– clay, loaded 5 kPa	14% dry weight (i.e. typical)	0.70	1.40
– clay, loaded 100 kPa	14% dry weight (i.e. typical)	1.20	0.83

clay binder has little effect. However, the thickness of these walls has a considerable effect on their insulating qualities and one would not construct walls of modern materials (bricks and blocks) to the same thickness. To allow a direct comparison to be made, U values are given for walls of the same thickness (220 mm) but constructed of different materials and it can be seen that earth is a better insulator than brickwork or stonework but not as efficient as lightweight blocks, due mainly to the interstices deliberately introduced into the material to improve its thermal performance.

It is interesting to compare a 600 mm thick earth wall with a modern cavity wall of brickwork and blockwork and note that the former has a lower (i.e. better) U value thereby confirming the warmth of the old mud cottage.

Other factors affecting insulation

Three other factors should be borne in mind when considering the insulating qualities of an earth building. Due to the nature of the

Table showing comparative heat losses through different types of walling

No.	Composition of wall	U value in W/m²K
1.	Brickwork	
	(a) 220 mm wall	2.26
	(b) 220 mm wall, 13 mm lightweight plaster	1.91
	(c) 335 mm wall, as above	1.51
	(d) 440 mm wall, as above	1.27
2.	Lightweight concrete blockwork	
	(a) 220 mm wall	0.74
	(b) 200 mm wall, 13 mm lightweight plaster and 13 mm external render	0.75
3.	Cavity walling	
	(a) 105 mm inner and outer brick leaves, 13 mm lightweight plaster	1.36
	(b) 105 mm outer brick leaf, 100 mm lightweight concrete block inner leaf, 13 mm lightweight plaster	0.92
	(c) 100 mm inner and outer leaves of lightweight concrete blocks, 13 mm lightweight plaster and 13 mm external render	0.59
4.	Chalk walls (10% moisture content)	
	(a) 300 mm wall, 13 mm lightweight plaster and 13 mm external render	1.12
	(b) 450 mm wall, as above	0.84
	(c) 600 mm wall, as above	0.67
5.	Clay walls (10% moisture content: 20% binder, 80% aggregate)	
	(a) 220 mm wall	1.59
	(b) 300 mm wall, 13 mm lightweight plaster and 13 mm external render	1.11
	(c) 450 mm wall, as above	0.83
	(d) 600 mm wall, as above	0.66
6.	Clay walls (10% moisture content: 30% binder, 70% aggregate)	
	(a) 300 mm wall, 13 mm lightweight plaster and 13 mm external render	1.16
	(b) 450 mm wall, as above	0.87
	(c) 600 mm wall, as above	0.69
7.	Stonework	
	(a) 220 mm limestone wall	3.06
	(b) 450 mm limestone wall, 13 mm lightweight plaster	1.78
	(c) 450 mm granite wall, as above	2.27
	(d) 450 mm sandstone wall, as above	1.65

material, the structural stability is maintained when openings are kept to a minimum both in number and size. In parts of Ireland, it was once common to erect a cottage without windows, the only opening being the door. Elsewhere, windows are small and restricted, thereby reducing the heat loss through the least insulated element of all.

The difference in nature of a building constructed of rammed earth compared to one of traditional construction cannot be emphasized too strongly. A semi-damp material rammed in thin layers with heavy rammers produces a wall almost devoid of interstices, whereas a wet material containing fibre, compacted lightly by comparison, produces a wall with a much higher percentage of interstices. The effect of those interstices is to increase the insulating qualities of the earth wall thereby making it superior to the wall of rammed earth.

Finally, the effect of moisture content should not be ignored. The higher the moisture content, the greater the thermal conductivity factor resulting in reduced thermal insulation.

THERMAL ADMITTANCE

The capacity of a material to absorb and store heat has been given added impetus in recent years in the search for energy efficiency. Electric radiators are available which allow heat to be absorbed during times of low demand and stored until required. They are charged during the night, encouraged by lower prices to level out fluctuations in demand. A heavy, dense form of firebrick is the usual medium for storing the heat and releasing it slowly throughout the following day.

The capacity of clay and chalk to absorb, store and release heat in a similar way has been known for some time and put to good use. Referring once again to S. Baring-Gould's writings of the west country, he comments that 'Cob walls for garden fruit are incomparable. They retain the warmth of the sun and give it out throughout the night'. In the chalk area of Hampshire, C. Vancouver writing in 1813 on the heat storage ability of a chalk garden wall stated that 'It preserves the young fruit and blossoms from the severity of the frost, and is thus recommended as affording a more certain crop of fruit which ripens as early, and is equally well flavoured as that (grown) upon stone or brick walls'.

Wessex contains many miles of chalk boundary walls which were built with the dual purpose of providing support for climbing fruit and security for the produce garden. This is witnessed by the horizontal rows of wires attached to hooks nailed into timber bearers cast into the wall on the sunny side. The walls are now old and the trees fully grown, but the truth of the comments by the nineteenth century writers can still be verified.

The ability of any material to absorb the radiant heat of the sun is understandable but the ability to store it is quite different and is largely dependent upon its density. Cast iron is a well-known example, hence its popularity for use as early radiators and fire backs. The density of chalk and clay differs little but varies slightly with the method of construction. Rammed earth buildings have a higher density and a greater heat storage capacity than those built by the traditional method which have higher volume of interstices. Although these improve its insulating qualities, they will not store heat.

Thermal admittance is defined as the rate at which heat will flow between the face of a wall and its interior. For practical purposes, the latter is taken as 100 mm deep as this is the limit to which the values are affected.

Thermal conductivity and resistivity factors used to measure heat losses can also be used to measure thermal admittance which is represented as Y values in W/m^2K. The higher the Y value, the lower its heat storage capacity but very thin materials such as glass have the same U value as Y value. Moisture has a fairly high thermal admittance value and a damp earth wall is capable of absorbing more heat than a dry one. In this respect, a damp wall is more advantageous than a dry one which is the opposite with regard to thermal conductivity.

Earth walls will only absorb heat when they are colder than the surrounding air. Conversely, they will only dissipate heat when the temperature of the surrounding air drops below that of the wall. The heat will flow from both radiant and convected sources through both faces of the wall. Where a building is heated, there will be a greater absorption through the outer face in the summer months and a greater absorption through the inner face in the winter months. Each rate of absorption will be different and there will be an insulating area in the centre of the wall isolating the two heat absorbing surfaces. Heat is dissipated back through the surface which absorbed it so that heat absorbed through the outer surface cannot be used to heat the interior of the building. This is why fruit trees are able to benefit when planted against the outer surface.

STRENGTH

Factors affecting strength

Design, moisture content, type of material, the addition of stabilizers, the method of construction, locality and quality of workmanship all affect the strength of a wall to some extent.

Little thought was put into the initial design, as the builders used rules of thumb guides gained from experience and passed on from one generation to the next. The thickness of a wall in relation to its height is a key factor together with the number of storeys to be supported in addition to the roof. The reason for battering the outer face, as is often found in the New Forest (Hampshire), is not really known but may affect the strength of the upper storey where the walls are thinner. There is an optimum thickness for a wall directly related to the strength of its material, and to exceed the optimum by building a thicker wall has no effect whatever.

It has already been mentioned that excess moisture added when building a wall will cause shrinkage cracks thereby reducing its strength. Therefore, the maximum strength could only be obtained if the optimum moisture contents previously stated for different materials in varying methods of construction were followed.

A wall constructed of unsuitable materials will last little time at all. The quality of the clay binder varies from one region to another, some are highly expansive such as those in south Hampshire and some, like those found in parts of north Devon, so remarkably stable that they need little or no fibre by way of additive. The aggregate can be examined closely with a magnifying glass to determine its shape. Angular grains interlocking together form few interstices and are preferable to spherical grains which have no natural bonding characteristics and leave large, irregular shaped interstices. Chalk provides the ideal medium, but when acting as an aggregate with a clay binder, the strength improves to that of the naturally found wychert which it closely resembles, probably because of a slight chemical reaction between the clay and the chalk.

The use of stabilizers make a considerable difference to the strength of a wall. It is doubtful if the addition of dung to a clay wall makes much difference as its primary function is to reduce plasticity, but its reaction with chalk may increase its strength slightly. The addition of hydraulic lime will ensure an early set whereas the addition of hydraulic or hydrated lime will increase the compressive strength. However, lime will have a greater effect on chalk than on clay because it will only react with the clay binder which is about 20 per cent of the mass, whereas when added to chalk it will react with the entire mass. The addition of cement has had limited use in the British Isles but a mere 5 per cent addition will increase the strength of the wall several times, rising considerably as its percentage increases.

The low moisture content and increased compaction of rammed earth construction ensures a high strength, particularly when chalk is used, as it is known to be well-suited to this method of building. At first floor level, walls were often built thinner but as they only need to support the roof, their strength probably varies little from those on the ground floor whose additional thickness is needed to take the weight of the upper floor. The strong, instant bond made when one chalk particle is rammed against another cannot be matched by the traditional method of construction where the initial bonding is weak, but gathers strength as the moisture reduces. Adobe blocks increase their strength during the drying out process and are not incorporated in the wall until full strength has been achieved. This is necessary to allow them to be placed in position without breaking up. However, the effect of the thick mortar joints cannot be ignored as lime mortar is considerably stronger than that made entirely of clay.

The effect of the locality on the strength of a wall is decided by the climate. Chalk is found in downland areas where the climate is relatively stable, but clay walls were once found in many parts of the British Isles with variable conditions. An area of low humidity will provide the best conditions for producing a clay wall of high strength and durability and it is these areas where they are still to be found.

The quality of workmanship is largely dependent upon the experience of the mud mason directing the operation as opposed to those executing the work. A binder:aggregate proportion of 20 per cent:80 per cent is ideal in most cases to provide a wall of maximum strength but the importance of a well graded aggregate of gravels, sands and silts to fill in the interstices cannot be overstressed. Adequate mixing, together with the careful addition of water, ensured that each grain of aggregate was thoroughly coated with clay. The builder needed to ensure that the previous course was dry before the next was added. He also needed to be aware of the limits of the material by not building a course higher than it could stand and ensuring that adequate compaction was achieved in relation to what the earth would allow.

Clearly, a number of variable factors decide the strength of a wall and a flexible approach is advised when using statistics obtained from one source and applying them to a different structure.

Weights and compressive strengths

Chalk

Chalk walls are constructed of Upper Chalk whose properties differ from Middle and Lower Chalks. Its weight varies depending upon its density and moisture content but it can weigh up to 2,400 kilogrammes per cubic metre (2.4 tonnes/m^3).

Once chalk is pulverized and reconstituted by the traditional method its crushing strength is reduced considerably. Compressive strength tests undertaken on test cubes prepared whilst a chalk wall was being rebuilt in 1983 failed at only 500 kilonewtons per square metre (500

kN/m^2). The wall was built by the traditional method, stabilized with a little lime and had a moisture content of 6 per cent.

Buildings of rammed chalk have a higher compressive strength than those built by the traditional method. The builders were aware of this and reduced the thickness of the walls without reducing the safety factor. The density achieved by compaction varied depending upon the moisture content, and a higher density was achieved if the moisture content was low. It is possible that a density approaching that of the natural rock could be obtained where the moisture content was carefully controlled. Cores taken from chalkroad embankments and tested by engineers have revealed an average density of 1.53 but as the range varied from 1.30 to 1.72 over the 3,000 samples taken it is reasonable to suppose that a similar range exists in walls of rammed chalk.

Chalk will consolidate itself under its own weight, even if loosely tipped, and after compaction it will continue to gain strength as the particles re-cement themselves together. The moisture content, however, will be much lower than that of the material rock due partly to the remoulding of the material but mainly to evaporation through the wall. This has the effect of increasing the compressive strength up to a maximum beyond which it decreases.

Clay
The strength of clay buildings varies more than chalk buildings because there is a wider range of materials involved. Clays, silts, sands and gravels vary both in percentage content and in character from one locality to another.

The weight of cohesive soils varies little due to their dense nature. Dry clay weighs 1750 kg/m^3 whereas silt weighs 1900 kg/m^3. However, cohesionless soils vary due to the volume of interstices which is determined by the shape and uniformity of the particles. Sands range from 1500 to 1700 kg/m^3 and gravels from 1300 to 1600 kg/m^3. By interpolation, a well graded clay mix with 20 per cent binder can vary in weight between 1600 and 1750 kg/m^3.

In a clay wall constructed by the traditional method, an average compressive strength of approximately 1000 kN/m^2 can be obtained if a sample is taken, dried and tested. In practice, the wall would not be completely dry and this would raise the compressive strength slightly. Even without any increase, this strength is safe for normal domestic loading.

Walls of rammed earth are considerably stronger. Compressive strengths of between 2100 and 2300 kN/m^2 have been obtained in tests on well constructed walls without fibre. However, if a little straw is added the results are increased slightly to between 2300 kN/m^2 and 2500 kN/m^2.

Clay lump construction varies from other methods in that their manufacture can be controlled more easily. The blocks are capable of being manufactured in advance in workshop conditions and not laid until each one has been inspected and approved. This preconstruction check, together with the bonding and mortar joints have a considerable effect on the strength of the building. Recently published results of compressive strength tests carried out on clay lumps show an average of 1500 kN/m^2 but occasionally reaching over 2100 kN/m^2. These tests have confirmed those made 70 years ago when a mimimum compressive strength of 1900 kN/m^2 was obtained.

Results of compressive strength tests are not directly comparable with other materials in common use. The unfired, perforated clay bricks used in an experiment at Bicton (Devon), had a compressive strength of under 2000 kN/m^2 compared to 37000 kN/m^2 for their fired counterparts, yet they were suitable for their purpose. Lightweight concrete blocks vary between 2800 and 3500 kN/m^2 and dense concrete blocks vary between 7000 and 10500 kN/m^2. These differences are counterbalanced by the thinner walls used in modern construction.

Tensile strength

Earth is a very weak tensile material and therefore prone to cracking. Tests taken on natural chalk have shown a variation of between 20 and 50 kN/m^2 with a moisture content of 23 per cent. However, a rammed chalk wall has a moisture content well below this and the effect will be to reduce its tensile strength to below these figures. Traditionally built chalk walls would be even lower and this stresses the importance of a well constructed underpin course and an evenly distributed load.

DURABILITY

Factors affecting durability

Durability is the ability to resist degradation enabling a building to last a long period of time. That so many earth buildings dating back several centuries are still standing is proof of the lasting qualities of the material which is determined by several factors. For an earth structure to be durable it needs to be suitably located, properly designed for its use, well constructed of suitable materials, adequately protected, used within its limitations and maintained regularly.

Location

Mention has already been made of the importance of location with particular reference to the strength of a wall. However, durability requires different criteria and the original builders chose their sites with care. As earth is capable of being remixed with water and reconstituted *ad infinitum*, it follows that its greatest threat is moisture. For this reason, earth buildings are not suitable in those regions where the climate is either wet or humid. In Ireland, none exist on the wet, west coast; in Scotland they are now found only on the dry east coast; in Wales and Cumbria they are found only in sheltered lowland areas and even in the damp south-western peninsular they are restricted to lowland areas or on the sheltered, leeward side of the high moorland masses. Lowland Britain is best suited to earth building.

Even in the drier, less humid areas, a dry site is essential, as it is desirable for all buildings. The absorbent nature of the material allows the walls to attract moisture by capillary action from the ground; a low moisture content in the wall is desired and a high one which fluctuates widely from one season to another is not. For this reason, a well drained, raised site is always to be preferred to a low lying, damp one.

The choice of aspect may play only a small part in the durability of a building but it should not be ignored. Geoffrey Grigson, a keen observer of buildings in the chalk downlands of Wessex commented on the need to erect buildings 'square to the compass . . . so that they present only a corner, not a full-faced wall, to the south-westerly winds', after noticing how many buildings which faced south-west had been damaged by weathering. When Dr Poore designed his chalk cottage at Andover (Hampshire) in 1901 as an experiment in modern sanitation, he set it out with the aid of a compass 'to ensure that there is a possibility of some sunshine upon every wall of the house at every season of the year'. The difference in heat absorption between south and north facing walls is considerable. The longer the south elevation the greater its heat absorption and thermal expansion thereby increasing the risk of subsequent cracking. This was often minimized by ensuring that the main elevation faced either east or west.

It should not be assumed that the temperature of an earth wall is constant at any given time. Variations occur between the outer surfaces and the centre of a wall as might be expected, as the centre is relatively stable because of the surrounding insulation from which the surface can claim no benefit. Another variation occurs between the outer and inner surfaces, the former is affected by wind, rain and sunshine causing regular fluctuations, and the latter only by the users of the building causing little fluctuation. A third variation is determined by height. The bottom of a wall is affected both by rising damp and maximum exposure to sunshine, each having an opposite effect on temperature. The same conditions are not found at the top of a wall, the eaves overhang ensuring almost constant shade and the reduced moisture content allows a lower rate of thermal conductivity.

Generally speaking, if a building can be positioned so that the radiant heat of the sun is absorbed and expended evenly, it will be to its advantage.

That the site is close to the source of the raw materials is fundamental, as two of the prime reasons for building in earth are the availability of suitable soils and the need to dispense with expensive transport arrangements. That it is free from disturbance by trees is also understood as growing roots cause subsoil upheavals which easily dislodge underpin courses resulting in uneven settlement and cracking of the walls.

The loadbearing capacity of the soil has a special relevance to earth buildings whose walls are usually twice the thickness of those built with modern materials. Earth is dense and heavy and the lack of footings to distribute the load over a greater area makes it even more essential to excavate down to a solid base in the subsoil. To build on filled land or on a site containing both excavated and filled land would result in a structure whose life was limited at the outset.

Design
When discussing the strength of walling, mention was made of the importance of design, and the same basic principles apply when considering durability. However, other factors also need to be taken into consideration.

The main aim was to avoid cracking caused by structural movement, far more important in an earth building with its comparatively low loadbearing capacity and low tensile strength than in one built of masonry. A building is designed for a purpose, and one would not expect a wall designed for a cottage to be capable of supporting the enormous roof timbers of a lofty barn. Wall thickness needed to vary with height and the number of floors being supported. The large, open plan of a barn requires stability at high level in the form of tie beams, whereas the small rooms of a low cottage rely on the bonding and integrity of the cross walls to provide the stability needed to the external walls. Each room forms a cell-like structure relying on its neighbour for support and achieving its strength like an egg crate or honeycomb.

In two storey buildings, the distribution of the dead load was carefully considered so as to support the roof on the end walls by means of purlins set on spreader plates, and the upper floor on the front and back walls by means of joists resting on wall plates. The ceiling and floor joists tie in the cross walls and the floor boarding acts as a diaphragm to strengthen the entire structure. In single storey structures, thought was given to distributing the roof load evenly over the four external walls. Purlins are essential both to tie in the flank walls and to make them take a load; tie beams are necessary to tie in the front and back walls. Spreader plates help to avoid point loads and reduce risk of cracking.

Openings need to be kept small to allow as large a mass as possible to take the load. Where formed, the lintel end bearings are much greater than in masonry construction and provision was made when the wall was built for fixing the joinery to avoid unnecessary holes having to be cut later. Windows were deeply set in the jambs and the quoins rounded to be able to take the odd knock without damage.

The siting and construction of the flue was of major importance. To locate it on an outside wall caused uneven loading thus inducing cracking either side, whereas to place it centrally in the building provided a strong core around which the floors and roof could be constructed, thus stiffening the entire structure. A brick flue is better able to cope with thermal expansion through its joints than an earth flue. It should be built quite independently of the rest of the structure without a bond between the brickwork and the earth walls, as lateral stability at this point was provided by tie beams or upper floor joists.

Materials and workmanship
The careful choice of suitable materials determine the durability of a wall above all other factors. The importance of a well graded gravel mix with the right balance of clay binder and straw has already been emphasized and does not need further comment. The variable characteristics of local clays have also been mentioned, some of which are more permeable than others. Where these are used, the expansion at the centre of the wall occurs later than that at the surface, resulting in slight movement. The fibre can normally cope with this but where the walls are very thick, structural cracking may occur.

Chalk is a stable material which expands and contracts little compared to clay. This is possibly due to the high calcium carbonate content which exceeds 95 per cent in southern England and which makes it so suitable for rammed construction. Chalks found in other parts of Britain have a much lower degree of purity which may account for the greatest concentratioin of chalk buildings being found in the south.

No further comment need be made regarding the standard of workmanship as long as it is realized that the raw materials require thorough mixing with the minimum amount of water and that the degree of compaction and the height to which the courses are built all help to determine the longevity of the structure.

Protection

Once completed, the building needed to be protected. The aim was to prevent excess moisture from entering the walls and destabilizing them. A small amount of moisture can be advantageous but all that is needed is absorbed from the atmosphere and nothing should be allowed to inhibit this.

The base requires the protection of an underpin course. Its function is not to distribute the weight of the wall over a large area like a foundation but to prevent erosion at this vulnerable spot. Its thickness is the same as the wall so that moisture running down the rendered surface drips clear. It acts as a breathing barrier between the wall and the ground allowing moisture to rise and evaporate on the outer surface of the joints, before reaching the earth. To assist it in this task it needs a ground slab higher than the outside ground level. It needs to breathe to allow the passage of ground moisture and to ensure that harmful salts are not deposited in the earth wall.

Moisture also attacks a wall from above, soaking into the earth, and causing it either to slump or be attacked by frost. Efficient overhead protection is necessary with a much greater than usual eaves overhang. Successive layers of thatch can build up to 600 mm giving a 750 mm wide overhang and ensuring protection, but such roofs do not have gutters and rainwater is shed well clear of the building. An absorbent surface needs to be provided around the building at this point as dripping rain will splash onto the wall just above the underpin course and cause erosion.

Low boundary walls with high underpin courses and wide eaves overhang usually receive no further protection, but a building is normally higher and it is necessary to protect the outer face of the wall. A thin, soft, breathing render is all that is required. It allows the wall to absorb what little moisture it needs from the atmosphere and to dispel it during warm, dry periods. Any other form of protection is likely to reduce durability.

Wind is a drying influence which helps to improve the durability of an earth building. When combined with rain, however, its effect is abrasive, concentrating on any weaknesses in the render, weakening them further by scouring action.

Limitation of use

A building is designed for a purpose and to ensure durability, it must be used within its limitations. Failure to observe this simple rule will cause stresses in the fabric with which the building will be unable to cope. The higher standard of modern living and the affluence it brings can call upon the building to cope with demands for which it was never designed, demands which can prove fatal.

There is a desire to increase floor space by converting the roof space into living accommodation. This adds an additional load to the walls which may be more than they can bear as loads are redistributed unevenly, risking cracking at the corners.

The modern trend to enlarge rooms by removing internal walls, weakening the 'eggcrate' design and removing the stability they provided to the external walls. The original builders realized the necessity of ensuring a strong bond between the internal and external walls and in some cases, they provided reinforcement in the form of wooden ties built into the walls between each course. To remove the cross tie is to allow each external wall to run wild unless stabilized by other means.

The wish to enlarge windows to admit more light causes damage at the top when longer lintels are fitted and reduces the overall mass of the wall, leaving less to take the load. Many small buildings are extended because they are not large enough for modern needs. Stability can be affected when a strongly built masonry extension is attached causing stress at bonding junctions and weakening the walls where new openings need to be cut to gain access. There is also a trend to convert a building designed for one use to another use. This can set up stresses from higher loadings and increase the loadbearing capacity of the walls to danger levels by the introduction of upper floors, partitions and bathrooms. This is relevant in the conversion of agricultural buildings to domestic use.

The effect of vibration on an earth building is a hazard of the current century and it is usually impossible to insulate against it. However, the surviving buildings tend to be in quieter rural areas and those that once stood alongside the main roads have fallen long ago.

Maintenance
The need for regular maintenance is basic to any type of building but more relevant to an earth structure. Lack of regular maintenance will result in repair work which can be more difficult to execute than in a masonry building. Once the protection has been punctured, the wall is exposed to rapid decay and the adage 'prevention is better than cure' cannot be stressed too strongly. Total lack of maintenance will result in the building being returned to the soil from which it was created.

Chapter Four

Repairs, alterations and extensions: general principles

INTRODUCTION

It is a mistake to alter, convert or extend any building whose character is adversely affected. Historical interest is usually reduced or sometimes lost and the charm which formed part of the integrity of the structure is altered in a way which can never be restored. This is often due to a change of setting where, for example, a church is converted to offices and the parish notice board is replaced with the corporate nameplate, blinds appear at the windows, fluorescent strip lighting floods onto the street and the graveyard is cleared of headstones to provide car parking facilities with lighting columns.

The need to alter, convert or extend arises as the wealth of the nation increases, and higher standards of living are demanded by the population. Small cottages built to house the large families of Victorian workers are no longer acceptable for the smaller families of the late twentieth century and an extension is considered essential to most prospective purchasers. Growing affluence demands more space. The shortage of accommodation coupled with the rise in the value of property has created the need to make the most economic use of the existing available space. This usually involves an alteration such as inserting an upper floor, or replanning the layout of the building by removing internal walls and flues. The rapid development of agriculture has left many farm buildings redundant creating a wealth of large, empty spaces in need of a new use if they are to survive. The financial pressures on the farming community, encouraged by enthusiastic developers, has led to considerable pressure on the planning authorities to approve applications for change of use. This usually involves the provision of amenities such as garages, conservatories, tarmacadam hardstandings and landscaped gardens, none of which are contemporary with the structure.

The threat to a building can be considerable. Apart from the loss of historical interest, the stability of the structure should be considered. The removal of a central flue, the insertion of an upper floor, the cutting of tie beams to form door openings, the enlargement of windows, the installation of a heating system, and the way in which the building is to be used, all affect its stability. The best advice is to repair and leave well alone for the good of the building. However, if it is considered absolutely essential that further work must take place, the advice of a structural engineer sympathetic to the care of older buildings should be sought before proceeding with any detailed design work.

Furthermore, it should be borne in mind that although repairs may be carried out without approval, any alteration, conversion or extension to any listed building which affects its character, or to any structure within the boundary of a conservation area will need Listed Building Consent and Building Regulation Approval from the local authority. Even non-listed buildings will probably require Planning Consent and Building Regulation Approval and enquiries should always be made to the local authority before commencing work on site.

THE PHILOSOPHY OF REPAIR

Most earth buildings are located in the countryside and were erected as simple, functional structures by and for poor, working class rural folk. They were built simply and cheaply and were poorly maintained on the premise that they could easily be ploughed back into the ground from which they had arisen and a new building could be constructed.

The philosophy has been superseded by a rise in affluence bringing about a change in ownership whose values are based on preservation rather than replacement. The need to borrow the finance necessary to purchase a building usually requires its retention until the debt has been repaid at the very least. Maintenance has therefore become more important and one need only observe a 'recently restored country cottage' to realize that the mark of the twentieth century has been placed upon it, both in the standard of craftsmanship and the introduction of modern materials. The effort is always well intentioned but often misguided, resulting in loss of character by the alteration or removal of original features, the addition of neo-vernacular features and inappropriate methods of repair. Efforts are made to ascertain what features might have been present and to reproduce them without any proof of their historical significance. Instead of being repaired, the entire structure is restored, the vernacular lost and the charm of the rural idyll replaced by the town dwellers view of modern country living.

The bulk of Britain's earth structures are farm buildings and cottages. Many of the former are now being converted into the latter, not just as a place in which to live but also as an investment and a tangible asset to pass on to the owner's heirs. This basic change in the approach to property ownership dictates that the investment shall be kept in good order with the result that cottages are better maintained today than at any other time in history. The introduction of listing status and the creation of conservation areas has done much to encourage owners to maintain their properties, not for the benefit of the individual, but for the heritage of the nation and its future generations.

The decision to erect a building carries with it an obligation to maintain it regularly to enable it to perform its function properly. Lack of regular maintenance leads to severe measures having to be taken to save a building from collapse and replacement. Such measures result in loss of character and historic detail and it is these features which need to be repaired or preserved with regular maintenance to enable the charm of a humble building to survive.

Sir Bernard Feilden has defined conservation as 'the action taken to prevent decay' (*RIBA Journal*, July 1990). If repair is necessary, then decay has already set in and it is too late to prevent it. Delaying conservation gives rise to repair and the two need to be considered together as there is no definable line when conservation becomes repair. Only the minimum amount of work necessary to repair a building should be carried out. Owners are easily led by work hungry contractors to execute more work than is really necessary with the result that a simple repair can become a major task involving the replacement of important elements and the consequential loss of historic interest. This is most commonly seen when external renders are replaced, rather than being patch repaired to match the existing finish. A soft lime render and its associated patina which follows the natural contours of an earth wall is lost when it is replaced with a hard, thick, heavy, dubbed out render. This presents a flat facade to the observer and naturally rounded corners are replaced with sharp arrises formed by stainless steel angle beads.

All repair work should be executed with a sympathetic approach. This is not easy, but is best achieved by careful choice of a specialist contractor who has knowledge of earth buildings and is able to work unsupervised, or one who shows an interest in the building, who is prepared to admit his lack of knowledge and seek and implement specialist advice. The contractor should be prepared to adapt to traditional building methods, seek further information willingly on materials and techniques not known to him and enjoy repair work of this nature. Specialist contractors exist throughout the country, usually based in large villages or small market towns. They tend to be small, local firms, often well established, with a good reputation, and with a family tradition of repairing vernacular buildings extending back over several generations. They employ most of the building trades directly in order to ensure control over the work, rather than rely on subcontract labour tendered on a lump sum basis down to a price. When this happens, control becomes remote. The employment of such a contractor may well limit competition but a single negotiated tender, although resulting in a slightly higher price, is preferable to working against the ticking clock of profitability as observation of the completed work will prove.

The preservation of the original craftsmanship is of paramount importance regardless of how poor it might be considered. It is a reflection on life in a small rural community at the time the building was constructed and its use as an historical document to researchers in social history should be respected. Its delight is often in its simplicity and to replace it with modern techniques will lose the very character which appealed to the owner upon purchase. Work of a high standard of craftsmanship should be observed carefully and used as an example of the standard to be achieved when repairs are being carried out. A lower standard should not be accepted as a token of respect to the original builder. An enthusiastic modern craftsman needs no prompting to accept this point and work can be ordered under his direction with confidence.

The introduction of modern materials should be treated with caution. A case for their incorporation in repair works might be considered provided that they are capable of being removed and replaced at a later date with traditional materials, should they fail to achieve their aim and provided that damage would not be caused to the building during the replacement operation. In other words, the introduction of modern materials should only be considered providing the decision is reversible and that the aesthetic appearance of the building is not impaired. Traditional materials are available from specialist suppliers and are to be recommended in preference to modern substitutes with claims of longevity, even though their lives may be shorter. The trend of using modern materials with a longer lifespan has been brought about by the high cost of labour and scaffolding, as it costs approximately the same to apply a decorative material with a long life as it does to apply one with a short life. In the past, only materials with a short life were available leaving the cottager with little choice other than to purchase them and probably apply them himself when he considered it necessary.

Earth buildings are not as strong as those of masonry construction and their fragility should be treated with respect. Those of rammed earth construction are considerably stronger and require little in the way of maintenance apart from redecoration every few years. Adobe buildings, constructed only of those blocks which were in sound condition after drying, have relatively few problems and regular re-decoration is the main item of maintenance. Mud and stud buildings tend to be of fairly weak construction and many have had a brick skin built around them to reduce the need for regular maintenance. Traditionally built earth buildings give the greatest cause for concern if maintenance has been neglected. In some parts of the country their very survival has reached crisis point and urgent action is necessary if the vernacular is to survive.

The reasons for decay have been discussed in Chapter Three and are dependent upon the location of the building, its design, the materials and workmanship, the degree of protection and maintenance it has been afforded and the way it has been used. Rain is the main enemy in the British Isles and all efforts need to be made to keep it at bay. Water rises into the wall from the ground and penetrates the wall from the top giving rise to the much quoted Devonian expression 'Gie un a gude hat and pair of butes, an' ee'l last forever'. Although boundary walls are seldom protected, the face of a building needs some degree of pro-tection. This point was acknowledged by Clough Williams-Ellis when he inserted into the Devonian expression 'and a good raincoat'.

All efforts must be made to prevent moisture from soaking into the wall, particularly from the top. The importance of a wider than average eaves overhang and well pointed ridges cannot be overstressed to prevent rainwater reaching a point where it will cause the wall to slump. Alternatively, on a very cold, clear, still night in the middle of the winter, the entire wall will fall without any warning and with a sound like a tremendous clap of thunder, razed by frost.

The risk of decay is greatest when the moisture content of a wall is at its maximum. It is a natural characteristic of the material to absorb moisture and its flow should not be hindered in any way. This ensures moisture absorbed from the atmosphere, or allowed to enter due to repair or lack of maintenance, can permeate, and evaporate on the surface. No attempt should be made to repair a wall until it has dried out enough to remain stable while the repair is being undertaken, or while there is the slightest risk of it being damaged by frost.

Delay in maintaining a building gives rise to repair. The more major the repairs, the less economic it becomes, threatening its very existence. Any building can be saved but often at high cost and with such consequential loss of historic detail that one should consider carefully whether it is worthy of retention. Rebuilding is the last option and should rarely be considered.

ALTERATIONS, CONVERSIONS AND EXTENSIONS

Alterations to existing buildings to form extra habitable rooms

Attic conversions
Provided there is adequate headroom, the temptation exists to convert the roof space into extra accommodation, although in houses built of rammed earth it is doubtful if there is sufficient headroom as roof coverings are usually of slates laid to a shallow pitch. It is important to consider all of the implications before making a decision as it may be better to leave the building intact and to extend it rather than apply stresses with which it is unable to cope.

If there is not sufficient headroom, one may be tempted to consider removing the roof, raising the walls and refixing the roof at the higher level. Such a proposal is expensive and usually alters the proportions of the building to such an extent that it is aesthetically unacceptable. It also affects the thickness to height ratio of the wall which may increase to a point where the wall becomes unstable requiring additional methods of lateral support. Removal of the roof will damage the top of the walls at eaves level and although this is capable of repair, a line of weakness is likely to remain. The effect of the increased loading on the foundations should be considered as the underpin course was only built to take the weight of the wall at its present height. Had the builders known that the height was likely to be increased, they would have built it more strongly with more through stones and perhaps with footings to ensure that the weight was spread over as large an area as possible. A trial hole is always necessary to ascertain the existence, size and construction of the underpin course.

The same points should be considered regarding the general principle of the provision of an extra storey. The building was designed for its present accommodation and the additional floor will involve both the extra dead weight of the structure and the imposed loads of the occupants and their furniture. The additional weight on the walls and underpin course may be more than they can bear and careful calculations are necessary to determine the margin of safety. One method which has been used to overcome the problem is to support the weight

of the floor on an independent structure. This has been achieved in
one of two ways. A new brick inner skin, constructed alongside the
existing loadbearing walls on independent foundations has the advantage of providing a flat, dry surface to receive the internal plaster but
the ground floor rooms are reduced in size and the character of the
rooms is lost. The alternative is to support the ends of the upper floor
joists on timber beams running alongside the external walls, supported
on brick piers in the corners of the rooms and built off independent
pad foundations. If such schemes are envisaged, it is important not to
transfer the weight of the roof onto the new structure as the existing
walls rely upon it for stability. An independent structure to support an
attic conversion in a dwelling is normally unacceptable and is not
recommended.

If calculations show that it is safe to allow the existing walls to
take the additional weight, thought should be given to its distribution.
It is important not to interfere unduly with the present distribution
between the front and back walls and the flank walls as a change
of stress may cause cracking at the corners with consequent loss of
bond. A simple roof structure ties the cross walls together and its
design should not be altered in any way which will cause the loss of the
tie and allow the walls to bow. A purlinned roof is designed on a
different principle with much of the weight being take on trusses
resting on the cross walls but with the flank walls providing end
support.

The re-siting of the water tank requires more thought in an earth
building than in a masonry structure for the very same reason. Its
weight should be distributed over as large an area as possible to avoid
a point load in one corner which can only result in a crack.

Any attic room will require daylight provided by windows in the
roof space. Some planning authorities consider it more aesthetically
acceptable to construct dormer windows than to install flush mounted
rooflights. Although their weight is not great, it is better to avoid point
loads by not mounting them on top of the wall, although if wall plates
have been used, their weight will be more evenly distributed. If it is
necessary to fit them, it is recommended that dormers are installed
within an opening framed within the rafters to enable the trimmer
joists to distribute the load over a larger area. Daylight is much better
provided by the installation of suitable windows in the gables.

Basement conversions
Earth buildings sometimes contain a basement, particularly in the
chalk areas of Wessex, formed by the removal of the chalk which was
used to build the walls. Although there might be sufficient headroom
to make use of the space, this was never the original intention. The

REPAIRS, ALTERATIONS AND EXTENSIONS: GENERAL PRINCIPLES

basement was boarded over with a suspended floor and provision was not always made for access.

The walls to the basement are usually built of small lumps of chalk laid at random in a soft, lime-based mortar, with the traditionally built or rammed chalk walls on top. Where the lump chalk rises above the ground level it acts as an underpin course but is usually rendered externally together with the walls.

The urge to turn a basement into a habitable room is tempting and has often been tried, but the implications should be considered in detail before any action is taken. If additional accommodation is essential, it is better to extend rather than alter and put the stability of the building at risk.

The basement walls seek lateral stability from the adjacent soil and careful internal inspection will reveal whether they remain stable or whether lateral ground pressure has interfered with them and induced bulging. They were never designed as freestanding walls and soil removal will increase the thickness to height ratio of the entire wall to such an extent that it would be in danger of collapse. A structural engineer can calculate the implication of the proposals.

If the building is terraced, removal of lateral support from each property individually until the entire elevation is exposed is very unwise and may result in claims from neighbours whose rights of support are inherent in the law.

Such large scale earthworks around a building cause alteration in the movement of ground water with unpredictable consequences. The formation of such a large sump can change the natural drainage pattern of the soil which can affect the stability of nearby property.

Local authorities usually require any proposal to alter a basement into a habitable room to be accompanied by calculations to show that the stability of its walls are not impaired following the loss caused by the removal of the soil. This often involves a scheme to increase the stability of the basement walls by the construction of an inner skin of brickwork. Such proposals should be reconsidered and alternative schemes prepared for submission to the authorities, as the construction of an inner skin combined with the removal of the soil will affect the moisture content of the wall and increase the risk of instability. The newly exposed wall will need protection, usually in the form of a render, to match that existing at the higher level but this is not easy to apply as the wall was built overhand and the face was unseen. Once exposed, the crudity of its construction can be assessed.

All habitable rooms require light and ventilation and if the basement is to have separate access, several openings need to be formed in the wall. These weaken the structure even further and must be taken into account when calculations are made for the proposals to stiffen the walls. The Building Regulations require that at least one twentieth of the floor area of a basement should be capable of opening for ventilation, some of which must be over 1750 mm above floor level.

The formation of a full height door opening and windows in a poorly built wall of low strength with its lateral support removed needs careful thought, but could be carried out with success where adequate precautions have been taken, although cracks must be expected in the render. Examples of altered buildings are shown in Figure 16.

Such alterations were never envisaged by the original builders. The symmetry of the elevations are thrown out of proportion, particularly

Figure 16 Making the most of the available space. Both of these adjacent buildings have had their attics converted to provide additional accommodation and lowered the ground levels to bring their basements into use, each with their own external access doors. Cracks have since appeared between the windows and the render has lost its bond with the earth wall in places.

in a terrace where only isolated properties have carried out the work, each with a different type and size of opening in various positions. The building is much better left as it was originally intended.

Change of use of agricultural buildings

Introduction

The pressure to convert redundant agricultural buildings to other uses has had a marked effect on the rural scene in recent years. Much local character has been lost and yet with a little thought, much could be saved. The general rule is to do as little as possible to a building in order to preserve that character which provided the initial attraction.

Most conversions are from agricultural to residential use but larger buildings are more suitable for commercial and light industrial use although economics may suggest that residential use is to be preferred in all cases.

There are a number of problems involved in converting a non-habitable building into a habitable one and the preservation of its character is one which is of paramount importance. Cottages and houses are always protected by a render but many agricultural buildings, on the other hand, have never been protected in this way and to render them will immediately extinguish all external signs of historical interest. The use of chalk slurries, mud renders, coal tar and other forms of protection local to the area should be maintained and the developer made aware of the need for regular commitment to maintenance costs.

A similar loss of character is usually apparent internally but this is largely unavoidable due to the need both to improve the insulation of the walls and to provide an acceptable, dust free surface. It is not wise to introduce a dry lining due to problems of providing a firm fixing to the walls and of sealing timber battens in a damp atmosphere, unless permanent through ventilation can be arranged. The introduction of an inner skin of brickwork will overcome these problems and provide a dry loadbearing structure upon which an introduced upper floor can rest but the air space behind the skin should be left open to allow air to circulate.

The cost implication of conversion should be studied in more detail than with other types of buildings and a healthy contingency sum held in reserve in preparation for the unknown. The repair of rat runs is expensive and the extent of the problem is not normally known until after financial budgets have been prepared. The extent to which ground levels need to be reduced, cracks repaired, falling ends rebuilt, leaning walls stabilized and underpin courses repointed is greater than may at first appear and these items are essential repair works necessary to put the building in good order before the cost of conversion can be estimated.

Once an agricultural building has been repaired and converted to another use, it is usually cared for much better than it was before. A successful conversion is one designed by a person who both understands the nature of the material with which he is working and who is sympathetic to the use of the building and its surroundings in its initial context. An unsympathetic design could result in loss of listed status.

Ancillary farm buildings

Stables, cow sheds, cart sheds, estate offices and outbuildings are often ideally suited for conversion to residential use. Their size is right, their spans are short, they were built with door and window openings and may be of one or two storeys in height. They were built like cottages and can easily be converted into habitable buildings with success, provided that a few general rules are observed to cater for the special nature of the material.

Maximum use should be made of the existing openings for doors and windows. The amount of alteration work to the walls should be kept to a minimum and the formation of all but essential new openings for doors and windows is discouraged.

All internal cross walls should be maintained for stability. The external walls rely upon them for support and they should not be removed. The walls were built as one homogeneous mass and need to be retained that way.

An unpleasant odour emanates from buildings where animals have been kept. Urine soaks into the walls but the smell is not always obvious until the building has been enclosed. It will eventually disappear but may take several years to do so after the animals have been removed. Plastering will not mask the smell but may retard it. It is better to ventilate the building until the smell has gone before plastering internally. Also, plaster adhesion may be a problem.

Stables are often two storey buildings and maximum use can be made of the upper floor as it is inherent in the structure. If headroom permits, an upper floor can be inserted in a single storey building, supported on an independent structure built within the external walls but not tied to it in any way. If the headroom is inadequate, it is not recommended that the ground floor be lowered and the walls underpinned, as the risk of cracking is great and the structural stability of the entire building is threatened.

Barns

Barns vary in size with the larger ones often being converted to light industrial or commercial use and the smaller ones to residential use.

They can be considered as halves of a building spaced apart but connected by a common roof with each half dependent upon the other for support. They have no internal walls and offer plenty of open space but this might be badly proportioned for comfortable living. Barns built with earth walls are usually smaller than those of timber framed construction and a residential use is usually preferred.

Their structural stability is seldom in doubt as they were built with substantial underpin courses and thick walls to take the weight of a heavy roof. A carefully considered conversion scheme can leave the stability intact or even enhanced. The major problems are usually to insert an upper floor and to bring light into the building without distracting from its external appearance.

There is usually adequate headroom to allow an upper floor to be inserted but it may be wise to provide this to only part of the area and to support it on an independent structure of loadbearing cross walls. No attempt should be made to bond new walls to the existing walls which are unaffected by the insertion and still support the roof. Together with the provision of a solid ground floor, a new structure will be formed within the barn giving it additional stability.

If an upper floor is provided to only part of the area it allows light to enter from the gables and from the full height openings in the centre of the building alongside the threshing floor. This allows adequate natural lighting to reach all parts of the building without the need to provide rooflights or to cut windows in the earth walls, although a certain amount of 'borrowed' light may be necessary. All existing openings should be utilized to dispense with the need to form new ones.

Barns offer a challenge to any designer but few of them are ideally suited for conversion to other uses. If a scheme involves the introduction of a chimney, the cutting of window openings, or the provision of rooflights, the building is not ideally suited for conversion and should be maintained until a suitable alternative use can be found.

Extensions to existing earth buildings

Most earth cottages are small buildings and therein lies much of their charm. However, the standard of living demanded by the cottage dweller of today demands more space, and an extension is preferred to overloading the fragile walls with the burden of an additional storey. The extension should be considered as a separate structure built alongside the existing building so that both are independent of each other. It should not be considered as a 'lean to' extension. Designs which involve an extension to join two buildings together to form one large building are seldom successful and historic interest is usually lost.

Local authorities usually require an extension to be built in a masonry type of construction, possibly rendered and decorated to match the existing building, but there is no reason why it should not be constructed of earth by the same method as that used originally. This should be considered and discussed with the authority as being a viable alternative.

The location of the extension should be considered with both aesthetics and structural stability in mind. It is seldom acceptable to extend at one or both ends to produce a long, narrow building. This not only impinges on the proportions of the original elevation obliterating the vernacular, but produces an unworkable layout. It is better to respect the front elevation and the role it plays in the local scene and to extend at the back to form a 'T' shape on plan. This allows the existing outline to be largely maintained and the extension to act as a buttress to stabilize any deficiency in the rear wall. For this reason, the two structures should not be bonded together and a loadbearing brick wall should be built alongside the existing external wall to support the upper floor joists and purlins of the extension. It will also provide strong fixings for heavy items such as kitchen and bathroom fittings. Any uneven settlement between the two structures due to differences in depths of foundation will cause no harm and allow them to move independently.

When planning the extension, the layout of the existing building should be respected and left unaltered. All internal walls, flues and staircases are better left in their original position, not just as a record of their historical significance but to assist the durability of a weak structure.

Part of the external wall will become an internal wall and access will need to be provided to the extension. Windows will need to be converted into door openings and this is preferable to cutting a new opening, as the delicate operation of lintel insertion is avoided. Any new opening will weaken the structure and should not be cut unless it is absolutely necessary. It is better to plan the extension around its access point based on an existing window rather than cut a new opening in the wall. Other windows are better left unaltered, but if an internal window is inconvenient, it can be filled in on one side of the wall to provide a feature.

No attempt should be made to copy the external elevations and a contrasting material, such as brickwork, will allow the extension to be evaluated on its on merits. An underpin course of, say, blue bricks, will allow the symmetry to be maintained even if the walls are to be built of earth. However, if they are constructed of blockwork and rendered, the undulating character of the original cannot be matched and should not be attempted. False features, such as rounded corners which are not appropriate to masonry construction, should be avoided. Storey heights, rooflines, degree of pitch, depth of eaves overhang and window proportions are better based on those existing but it should be obvious that the extension is not part of the original building. Any attempt to deceive will reduce its merits.

The importance of protecting an earth building has already been considered and it is at a vulnerable stage during extension. Temporary covering to prevent rainwater entering exposed wall heads at eaves level or the jambs of newly formed openings are more essential with an earth building than with one of masonry construction.

If the occasion arises where a new brick wall meets an existing earth wall in the same plane, no attempt should be made to block bond the two together. The loose end of the earth wall is better stabilized by a 'U' shaped support built in brickwork like the end of an open spanner, thus allowing each wall to move independently but offering lateral support.

When extending an earth building, the rule as always should be to interfere with the structure as little as possible, to respect its comparatively low strength and to do nothing which will impair its durability.

ANALYSING SOILS AND WALL SAMPLES

Before attempting to repair a building, it is advisable to ascertain the materials used for its construction. In some instances, it is possible to reuse the existing materials, though this is not always wise, and it becomes necessary to match them, particularly for colour, if the walls are exposed.

Various rule of thumb guides to determine the suitability of subsoils for building have been used. A simple test can be made on site after a shower of rain by kneading a handful of subsoil into a ball and observing its cohesion when released. If the ball retains its shape for one minute, it is worthy of trial. An eighteenth century test involved removing large stones and vegetable matter and ramming the soil in layers in a tub with splayed sides. Once full, the tub was turned over and removed to expose a mound of compacted soil which was inspected daily to ascertain cracking, crumbling and shrinkage as the moisture dried out. The soil was considered fit for building if these faults were not present. A nineteenth century guide records soils as being suitable for building 'if a pickaxe, spade or plough brings up large lumps at a time, if arable land lies in clods or lumps, or if field-mice have made themselves subterraneous passages in the earth'. It was also considered suitable 'when the roads of a village, having been worn away by the water continually running through them, are lower than the other lands, and the sides of the road support themselves almost upright'. Or similarly, 'when kneading with ones fingers the little clods of earth and finding a difficulty in doing it or . . . wherever there are deep ruts on a road'.

Such guidelines have obviously played their part in the development of the vernacular, but, although scientific analysis is seldom necessary today, simple sedimentary tests can be made to determine the composition of a wall for matching purposes. Walls of pure chalk do not warrant analysis as they consist of only one material but clay bound walls can be analysed to give an indication of their content.

If possible, take two samples from each course, one from the weathered face and one from the unweathered face as there can be considerable differences between them. Put each sample into a separate clear, cylindrical container, fill with water, shake vigorously and allow to settle. The heaviest soils will settle at the bottom and the lightest at the top, the former settling quickly and the latter slowly. Once the water has cleared, the fibre will float on top and the approximate percentages of gravels, sands, silt and clay can be measured. Yellow-brown coloured water indicates the presence of dung and if the breakdown is inhibited in any way, the presence of lime must be suspected.

Close examination with a magnifying glass will determine whether or not the sands are spherical or angular. If a more accurate analysis of the sands and gravels is required, this will have to be undertaken separately using sieves in accordance with British Standards Specification 1377 Part 1 (1990) 'Soils for Civil Engineering Purposes'. This will differentiate between fine (0.06 to 0.2 mm), medium (0.2 to 0.6 mm) and coarse (0.6 to 2.0 mm) sands, and fine (2 to 6 mm), medium (6 to 20 mm) and coarse (20 to 60 mm) gravels. The same method can be used for chalk samples if necessary.

Knowledge gained by analysing many samples can be summarized and plotted in graphic form to show the range of soils used to construct earth walls and for comparison with new samples, as shown in Figure 17.

More simple tests can be made in the field by dampening each sample, kneading it, flexing it between the fingers, biting it and examining it to form an opinion, but this is the skill of the soils engineer and requires considerable experience. Further guidance is contained in British Standards Specification 5930 (1981) Table 6, but unless experience has been gained by tests made on site which have been followed up with wet analyses, it is not recommended for use by the untrained.

Other simple tests can be undertaken to determine the binder: aggregate ratio by drying the sample, weighing it, washing out the

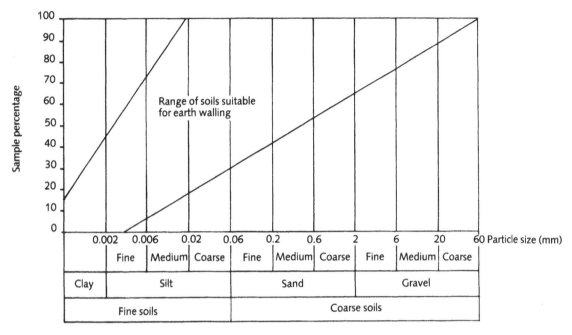

Figure 17 Range of soils suitable for earth walling.

binder in a bucket with plenty of running water until it is clear, drying it and weighing it again. Comparison with the original weight will determine the weight of the clay which has been lost.

Having analysed a sample from a wall it is necessary to match it from local sources. Subsoil can be excavated and analysed in the same way and the results compared to show whether the subsoil is deficient in cohesive clay and silts or in cohesionless sands and gravels. The deficiency can then be corrected.

These tests are simple and can be carried out without any previous experience, but scientific analyses are available from specialist laboratories which can be employed to determine mineral content, particle size, alkalinity, salt content, expansiveness of clay binder, porosity, thermal expansion, compaction, strength, adhesion, moisture content and plasticity, etc. should this information be required.

Chapter Five

Repairs, alterations and extensions: methods

THE UNDERPIN COURSE

Most buildings are constructed off underpin courses built of stone, flint or brickwork. Although the occasional building may be found which appears to have been built without one, a few minutes work with a spade usually reveals that the ground levels have been raised to cover it and allow the damp earth into direct contact with the wall. This is particularly common where a path has been laid around a building without it having first been excavated. The moisture in the ground soaks into the wall and rises by capillary action carrying with it nutrients in the form of harmful salts. Surface tension attracts the moisture and nutrients to the face of the wall where the moisture evaporates, leaving the salts behind, on the surface and also within the wall. Frost attacks the moisture near the outer face causing it to expand upon thawing, bursting off the surface and exposing the inner core, making it vulnerable to a further attack. This action must be prevented as the weight of the structure is being taken on a wall of ever reducing thickness which will eventually lead to its collapse. If the underpin course is covered by soil it is not performing its function and action must be taken to rectify the matter.

Occasionally, buildings are found without underpin courses. At one time they were probably more common but many must have fallen because of the absence of underpin courses.

To reduce the ground levels around a building is a simple matter but should not be undertaken unless trial holes have been excavated at regular intervals to ascertain the depth of the underpin course. This enables the depth of reduction to be determined. The underside of the underpin course needs protection against frost so an adequate depth of cover must remain. The ground levels can be reduced, provided it is safe to do so, to ensure that the floor level is approximately 150 mm above the ground level. This action will allow the moisture content of the wall to reduce, enabling any erosion to be repaired.

The newly exposed underpin course may well have roots growing through it. These need to be killed by a systemic weedkiller before being cut off flush with a pair of secateurs to ensure that there is no regeneration. The face of the underpin course may need to be repointed. This should be done by raking out decayed mortar, finely spraying with water to remove debris and repointing to a depth of at least 25 mm, or twice the thickness of the joint, with a weak lime mortar composed of one part of lime putty to three parts of sharp sand, well packed and pointed to match the existing wall. The exact depth of repointing is determined by the size of the stones and the degree of mortar decay.

The action of lowering the ground levels to expose the underpin course allows rising ground moisture to evaporate on its face before reaching the earth wall, thereby keeping it as dry as possible, to the benefit of the users of the building. It is the all important filter, isolating the wall from the soil and it needs to breathe through its joints to function properly. If it has been rendered with a hard cement-based mortar together with the wall, rising moisture is inhibited from evaporating before reaching the wall and the render should be gently chipped away from the face of the underpin course and the joints repointed. However, a traditional, soft, lime-based mortar will add to the filter and cause no harm.

Air needs to circulate freely around the underpin course and if it is not convenient to rake away the soil, well clear of the building, then a narrow trench may suffice. Once dug, it is wise either to slope the outer face or to construct a low retaining wall to keep the trench free from falling debris otherwise the problem of rising damp will return. If it is considered absolutely essential that the original ground levels are retained, then it will become necessary to backfill the trench with large angular stones to ensure that air circulates freely around the stones as far as possible. Earth will eventually clog the stones but moisture can be prevented from passing into the wall if a porous membrane such as 'Terram' is placed against it. However, these are unsatisfactory arrangements and should be avoided if at all possible.

Wherever space allows, the reduction in ground levels should be as wide as possible with the surface finished to slope away from the building. Where this is impractical and a trench has to be dug, one should consider installing a land drain alongside, at lower level, to channel away all surplus moisture which might otherwise collect in the trench and soak into the underpin course. The porosity of the soil will determine the need for this: chalk and sandy soils are free draining but clay soils are moisture retentive. The site should be kept as dry as possible at all times.

REPAIRS TO THE WALL

Introduction

Repairs are necessary if decay has set in. This is usually due either to action by the weather coupled with lack of regular maintenance, or interference by animals or vegetation.

Much can be learned by examining old buildings which have been repaired some time ago and assessing how successful the repair has been. Such repairs should be noted for possible future use, and several recorded by the author are reproduced for consideration. It is unfortunate that those repairs found most commonly are those of comparatively recent origin and involving the introduction of modern materials. The pointing of small cracks with cement-based mortars, the pugging of large fissures with concrete, the removal of soft spots and piecing-in with bricks, the cutting out of eroded areas and replacing with concrete blocks, or the introduction of metal lathing nailed to a wall prior to rendering, are all examples of recent practices which are not necessary. These can cause long term problems and should be avoided. The possibility of removing modern materials and replacing them with traditional ones should always be explored, but removal can cause further damage and it may be better not to put the stability of the building at risk. Each case therefore needs to be considered on its own merits.

Many rush to repair when it is not really necessary to do so. Property owners can become concerned at eroded areas, although repair is not always necessary. The location of the decay in relation to the load-bearing function of the wall at that point should be considered when deciding whether or not a threat exists. If a threat does not exist, there is no point in stressing the wall and putting it at risk by repairing it. It may be more sensible to stabilize the wall surface to halt the erosion and to prevent further deterioration. Tensile stresses built up in a wall are released instantly when a crack appears and it may be better not to stress the wall again by repairing it now that the threat has passed. Such a decision would depend upon the position of the crack, the type of structure in which it has occurred and whether or not any further threat exists. Only carry out a repair if it is necessary to do so.

Earth dwellings have usually been protected externally by a coat of render but farm buildings and boundary walls have often not received any form of protection except in East Anglia where tar has been used on clay lump. In recent years it has become common practice to render any earth structure to increase its durability. This action has had a detrimental effect on the environment by removing the patina and robbing the region of its individual character. Serious thought should be given to maintaining the distinctive appearance of a structure rather than obliterating it and creating the monotony of a national vernacular.

Low level erosion

Erosion to the surface of the wall just above the underpin course (Figure 18) is caused by excess moisture, either by rising damp as previously discussed or by rainwater dripping from thatched eaves onto a hard surface and splashing onto the wall. A wall with a high underpin course is protected from this problem but it is wise to lay adjacent paths with absorbent surfaces such as gravel rather than hard surfaces such as concrete which can only hasten the decay of the wall. Although it is the action

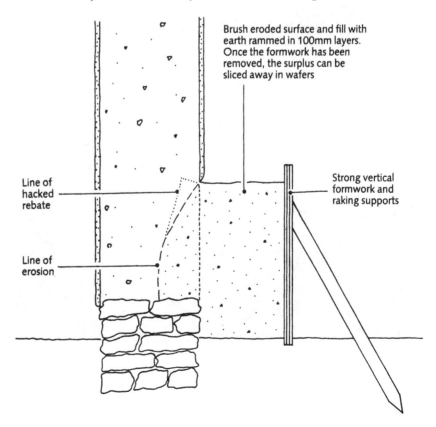

Brush eroded surface and fill with earth rammed in 100mm layers. Once the formwork has been removed, the surplus can be sliced away in wafers

Line of hacked rebate

Line of erosion

Strong vertical formwork and raking supports

Figure 18 Repairing erosion at low level.

of frost which bursts the surface initially this is exacerbated by wind causing a slight hollowing, which can deepen rapidly if action is not taken.

If the ground levels are high (see Figure 19), they should be reduced to a level below the top of the underpin course to provide a sound working platform. The eroded areas should be protected and the wall allowed to dry out before attempting to repair it. Using a scutch hammer, hack the top of the eroded edge to provide a distinct undercut horizontal rebate, and retain the earth for analysis. A scutch hammer is ideal for cutting earth walling as its replaceable, serrated blades allow the wall to be hacked without risk of cracking it. It may be possible to use an electrically operated disc cutter but if the wall contains flints or other sharp stones its use may be dangerous to the operator. On no account should a hammer and bolster be used as this will crack the wall, breaking the homogeneous mass upon which the earth relies for its stability.

Figure 19 High ground levels cover the underpin course allowing moisture to soak into the earth, and the hard, heavy, thick render prevents the wall from breathing. Once the moisture content of the wall reaches its liquid limit, it slumps.

Strong vertical formwork should be erected parallel to the wall, supported by raking timbers driven into the ground. Its height needs to be level with the top of the rebate. Having analysed a sample of the earth, it should be matched as far as possible from local sources. Salvaged earth and fallen debris are not recommended for reuse as the former will contain accumulated salts caused by rising damp, and the latter micro-organisms absorbed by contact with the soil. Any salts present in the fallen debris will probably have been washed out by rain together with the clay binder. The local subsoil may prove to be adequate with the addition of a little sand or clay.

As the base of the wall is highly vulnerable to low level erosion, it may be advisable to stabilize the repair by a 10 per cent addition of lime putty to the earth. This will not affect the colour of chalk or wychert but will lighten a clay wall slightly. The soil, sand and lime should be mixed and a sample squeezed in the hand to test for adhesion. If water is required, only the very minimum needs to be added. Due to the low moisture content, shrinkage will be negligible and it is not normally necessary to add fibre.

The inclusion of lime putty has the effect of strengthening the mix and its use is only recommended if the bottom of the wall is considered to be prone to further wear.

Having brushed the face of the wall to remove loose material, the earth should be thrown into the shuttering in 100 mm layers until it is filled. Each layer is rammed thoroughly to ensure that it adheres to the face of the wall, particularly the top layer to ensure that the rebate is properly filled. The formwork can then be removed immediately and the 'foot' sliced away in vertical wafers, using a sharpened spade, until it is flush with the face of the underpin course. The rammed earth will be very hard, similar to a soft rock, particularly if chalk has been used, and considerable effort is needed to slice it. To make the task easier ensure the spade is freshly sharpened.

It may be necessary to remove the spade marks and to indent the surface slightly to match the surrounding area and to make the repair less obvious. This can be done using a small pick, and a slight horizontal indentation hacked between the existing lift lines will make the repair almost unnoticeable.

If lime putty has been added, the repair may shrink slightly as it dries and the addition of a little fibre is recommended to avoid cracking. It will only dry if in contact with air and if the repair is deep, the centre of the wall may never dry out completely, and it will become progressively harder nearer the surface. If a slight crack should appear around the edge of the repair, this can be cut out and grouted in lime putty to prevent moisture from entering.

Having completed the repair, the ground levels can be lowered further to prevent a repetition of the problem. To have lowered them earlier would have resulted in abortive work, involving higher form-work and additional filling in excess of requirements.

High level erosion

Erosion to the surface of the wall at a higher level is usually caused by inadequate eaves overhang, rainwater soaking into the top of the wall due to inadequate protection, animal activity or vegetation (Figure 20).

Figure 20 Erosion to the entire face of a wall caused by an inadequate eaves overhang is exacerbated by low level erosion due to dripping rainwater splashing off a hard paving. Vegetation soon invades any horizontal ledge causing further erosion as the roots develop.

The importance of a wider than usual eaves overhang has already been mentioned. If it is narrow, dripping rainwater is blown onto the surface of the wall at high level where it soaks in, is attacked by frost and the resulting damage exacerbated by wind. Lack of suitable protection to the top of the wall will allow rain to enter and soak into the earth until it can take no more, eventually slumping (see Figure 21). The importance of replacing dislodged or missing tiles, of pointing the ridges and of re-thatching before the fixing spars are visible is therefore apparent.

Saturation of walls caused by blocked rainwater heads causes damage to masonry buildings and is a frequent initial cause of dry rot in structural timbers. In earth buildings the problem is more acute. If rainwater is unable to drain away freely, it will overflow and run down the face of the wall and the outside of the pipe. It is therefore a simple but wise precaution to ensure that the rainwater head is not fixed flush to the face of the wall and that the downpipes are fixed on long distance pieces.

Figure 21 Failure to maintain the thatched coping allows rainwater to soak into the top of the wall which freezes in cold weather and expands upon thawing, bursting off the render and taking part of the top of the wall with it.

Bushes, close to a wall, rocked by the wind, will erode its surface, deteriorating it further as they grow. Vegetation prevents surface tension from drying out a wall, keeping it damper than it needs to be and encouraging the growth of algae. It is better to remove the bushes to ensure that the wall dries out and the algae die, rather than treat it with a biocide which would only offer a short term solution.

Cows will erode the face of a wall by licking it and by brushing against it. Horses can inflict similar damage but builders often provided for this in their design by providing horizontal rails around a stable fixed to timber blockings cast into the wall and by providing brick jambs to all doorways through which the animals were to pass.

The masonry bee (*Osmia Rufa*) will attack the face of an unrendered wall by forming a mass of small concave indentations, with smooth, shell-like linings leaving only a thin wall of earth between each one (Figure 22). This erodes, but the insects are persistent and will form

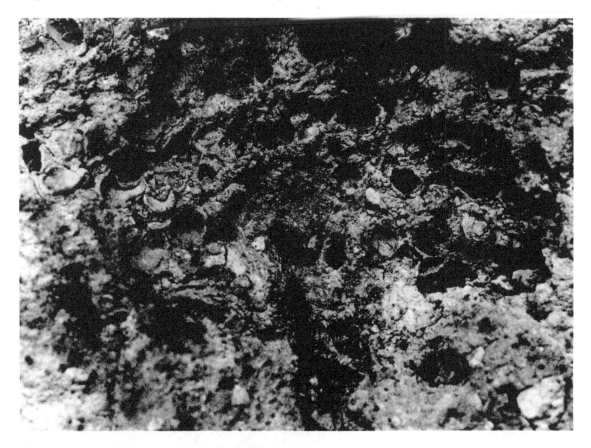

Figure 22 The masonry bee (*Osmia Rufa*) will attack the face of a wall to build a nest in which to lay its larvae. Damage is usually slight but a large colony will cause the surface to erode over a number of years. The bee is a persistent insect and is not easily deterred by attempts to discourage it from nesting.

new indentations alongside breaking down the surface still more. Prevention is better than cure and at the first sign of attack, the bees may be persuaded to seek an alternative home if fine mesh is hung in front of the wall while they are at the active stage in spring. Rendering provides a permanent solution, but may not always be appropriate.

If the colony is well established, the problem becomes more difficult to overcome. The insect bores into the wall in the spring, lays its eggs, provides pollen for the larvae and seals each cell with mud. The bees emerge in the summer but continue to use the cells until the following spring, when the cycle is repeated. The cells are therefore inhabited permanently. They can be filled, but the persistent nature of the insect will guarantee that new cells will be formed alongside to take their place.

The damage to the wall is seldom great and unless the bees are a nuisance it is best to leave them alone. If drastic action is really necessary, each cell should be injected in the spring with Permethrin dissolved in water or the entire wall sprayed each autumn for several years until they have been eradicated.

Repairing high level erosion requires techniques totally different to those used when repairing low level erosion. Where the latter comprised a method of ramming within a confined space, the former has no such limitation, indicating that different methods can be adopted. As damp does not usually rise to more than one metre, there is little likelihood of salts being present in the wall, so salvaged earth can be reused with confidence. Furthermore, being in a less vulnerable position, the addition of lime putty is not usually necessary.

The horizontal lift lines between the courses are vulnerable to erosion and they need special care if the character of the wall is to be maintained when the surface is not to be rendered (Figure 23). Using a scutch hammer, a horizontal chase at least 100 mm deep should be cut at lift line level, its height determined by the extremities of the erosion in successive courses. The upper surface of the chase should be under-cut to form a dovetail in cross-section. Hazel spars are obtainable from the thatcher and can be tapped gently into pre-drilled holes at 100 mm staggered centres, on skew, both above and below the lift line to obtain a secure fixing. Using secateurs, cut off the spars level with the face of the wall but at a sharp angle. Brush out the chase to remove debris and spray with water to ensure that the wall is wet but not saturated. A little clay and sand can then be added to the salvaged earth together with fibre to match that in the wall and the whole mixed thoroughly together with a small amount of water to produce a very stiff mixture. The chase is packed with the earth ensuring that the corners of the dovetail are filled and that the repair stands slightly proud of the surface. After a few days, the excess can be hacked back gently using a

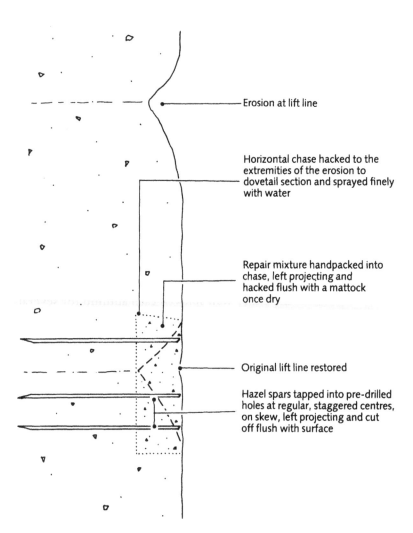

Erosion at lift line

Horizontal chase hacked to the extremities of the erosion to dovetail section and sprayed finely with water

Repair mixture handpacked into chase, left projecting and hacked flush with a mattock once dry

Original lift line restored

Hazel spars tapped into pre-drilled holes at regular, staggered centres, on skew, left projecting and cut off flush with surface

Figure 23 Repairing erosion to the lift lines.

mattock to finish flush with the existing wall and the position of the original lift line restored using a hand pick.

The same principle can be used to repair small or even large areas of shallow erosion or where the top of a wall has slumped due to saturation caused by a missing coping (Figure 24). Hazel spars are tapped into a grid of pre-drilled holes, on skew, until firmly fixed and cut off flush with the face of the wall. A firm, horizontal bearing needs to be formed at the base of the eroded area by hacking out a rebate with a scutch hammer to the depth of the erosion. A similar rebate then needs to be hacked at the top before the recess is brushed free of debris and sprayed with water to ensure adhesion.

An earth mixture to match the original should be prepared to which fibre and water are added and mixed thoroughly to form a very sticky

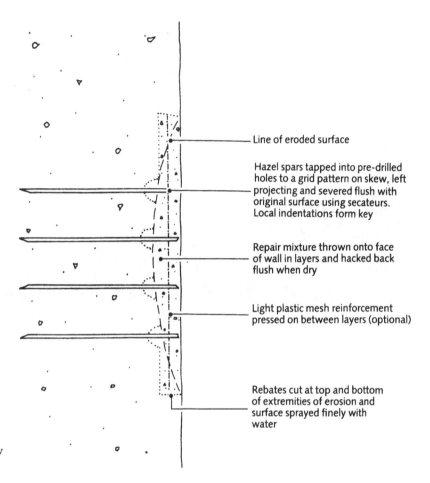

Line of eroded surface

Hazel spars tapped into pre-drilled holes to a grid pattern on skew, left projecting and severed flush with original surface using secateurs. Local indentations form key

Repair mixture thrown onto face of wall in layers and hacked back flush when dry

Light plastic mesh reinforcement pressed on between layers (optional)

Rebates cut at top and bottom of extremities of erosion and surface sprayed finely with water

Figure 24 Repairing shallow erosion at high level.

mass. The material can be thrown onto the face of the wall to form an instant bond with the eroded area and the spars. Should there be any difficulty in achieving a bond, a little lime putty may be added and mixed well in. Depending upon the depth to be filled, it may be advantageous to reinforce the repair lightly by pressing in a piece of light duty plastic mesh carefully threaded over the spars, before throwing on the final coat and finishing slightly proud of the surface. Once dry, the repair can then be finished with a sharp mattock and the original lift lines picked out if desired.

This simple but effective method of repairing a shallowly eroded surface can also be used to fill in isolated indentations where only one hazel spar anchor is needed.

If the eroded area is fairly deep, one of several methods of repair should be selected.

By far the most commonly found repair comprises chopping out the eroded surface to a depth of half a brick and filling in the recess with

Figure 25 Surface erosion has been 'repaired' using brickwork and the original soft lime render has been lost due to patch repairs carried out using cement. Being harder than the wall, rainwater runs off causing erosion to undamaged areas leaving the repaired areas standing proud. Note the low level erosion due to raising the ground levels above the top of the underpin course.

bricks or concrete blocks bedded in mortar to finish flush with the face of the wall. This repair is much stronger than the wall and allows rain to run down the face of the brickwork and erode its bearing, eventually leading to its collapse (see Figure 25). The repair has therefore compounded the problem instead of solving it. It is, of course, extremely unsightly but even if rendered can still cause problems. The bricks or blocks have a density and porosity different to that of the wall and greater coefficients of expansion. Differential movement caused by solar heat absorption forms cracks in the render through which rain will pass. This will result in loss of adhesion of the render coupled with erosion around the repaired area. It should therefore be borne in mind that a repair made with such materials cannot be considered permanent and that the resultant task will be much larger than the original one.

However, the principle of using precast earth blocks to repair areas of deep erosion should always be considered as the first option (Figure 26). This is particularly the case if the repair is to an eroded area which takes a point load, such as the end bearing of a lintel or a purlin in a half-hipped or gabled roof.

To effect such a repair, a scutch hammer should be used to hack out a recess at least 100 mm deep to the minimum rectangle necessary to enable blocks of a predetermined module to be inserted. The base of

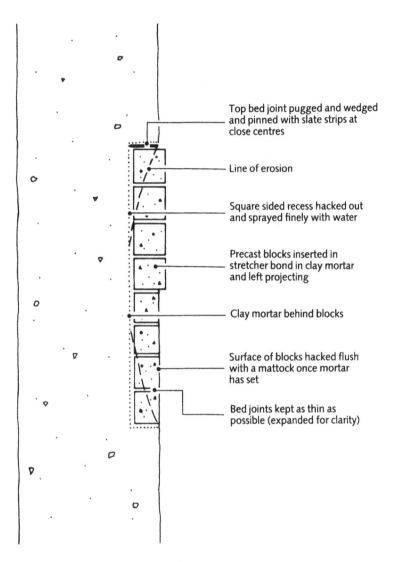

Top bed joint pugged and wedged and pinned with slate strips at close centres

Line of erosion

Square sided recess hacked out and sprayed finely with water

Precast blocks inserted in stretcher bond in clay mortar and left projecting

Clay mortar behind blocks

Surface of blocks hacked flush with a mattock once mortar has set

Bed joints kept as thin as possible (expanded for clarity)

Figure 26 Repairing erosion using precast earth blocks.

the recess should be level and the earth can be retained for reuse but a little clay and sand may need to be added together with fibre and just sufficient water to form a stiff mixture. A wooden mould comprising four sides and two lifting handles should be constructed, wetted and placed on the ground into which the repair mixture is thrown, struck flush with the top and carefully lifted off. A second mould of the same width and depth also needs to be made but whose length is only half that of the first mould. More blocks should be made than are required to allow for losses during the drying process. The blocks can be left under cover for several days before being turned and left for several weeks before the moisture content has reduced sufficiently for them to be handled safely without risk of breakage. When dry enough for use, a certain amount of shrinkage will have taken place.

The main advantage of this method of repair is that most of the shrinkage takes place before the blocks are incorporated in the wall, smaller blocks shrinking less than larger blocks. Small blocks are also easier to handle and lighter to place in position but any which contain cracks should be rejected.

The recess may need to be enlarged slightly to ensure that the blocks fit neatly, adjusting the height to allow just a few millimetres for each bed joint, plus an extra joint at the top for wedging and pinning. The length should be adjusted to allow the same thickness of perpend as the bed joint plus an extra one at the end. The depth should be equal to the finished thickness of the blocks. Brush the recess free of debris and lay a bed of clay and sand mortar, pre-wetting with a fine spray of water if necessary. The temptation to increase the strength of the mortar should be overcome as it is wise to ensure that it is weaker than the blocks. Additives such as lime or cement are not recommended. The first row of blocks are inserted with mortar perpends and mortar filling at the back to allow the blocks to stand proud of the surface. The second course of blocks can then be laid in stretcher bond utilizing the half-sized blocks at either end to break the vertical joint. The third course will match the first and the fourth will match the second until the recess has been filled leaving only a narrow horizontal gap at the top.

Wedging and pinning must be executed very carefully for fear of cracking the wall, and no attempt should be made until the mortar is dry. The gap should then be pugged solid with mortar and narrow strips of slate, whose length is equal to the width of the recess, tapped in gently at close centres along the joint. Shrinkage will be at a minimum if the pug is a mix of clay and sand to which the minimum of water has been added to ensure plasticity. Once the wall is dry, the face of the blocks can be hacked back flush with the wall using a mattock.

If a chalk wall is being repaired by this method, the blocks can be cast from chalk, fibre and water and laid in a mortar of chalk and water. Pugging for the final bed joint will have to be mixed more thickly.

An alternative method of repairing areas of deep erosion can be used, but it is rather slow and requires extra care as the wall is weakened during the process. It is an *in situ* repair but has the advantage of being tied back to the original structure (Figure 27).

Using a scutch hammer, horizontal grooves, a minimum of 100 mm deep, should be hacked at the top and bottom extremities of the eroded area, the upper should be undercut to form a dovetail in cross-section.

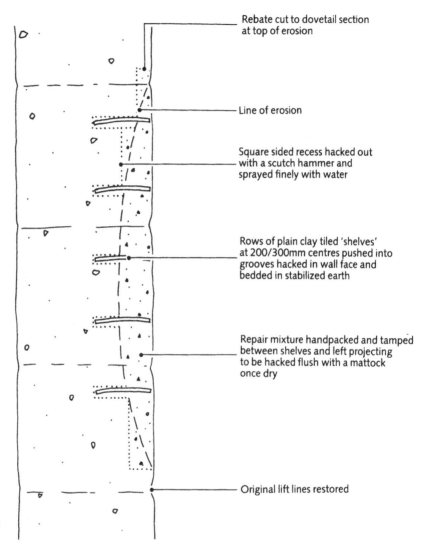

Rebate cut to dovetail section
at top of erosion

Line of erosion

Square sided recess hacked out
with a scutch hammer and
sprayed finely with water

Rows of plain clay tiled 'shelves'
at 200/300mm centres pushed into
grooves hacked in wall face and
bedded in stabilized earth

Repair mixture handpacked and tamped
between shelves and left projecting
to be hacked flush with a mattock
once dry

Original lift lines restored

Figure 27 Repairing erosion
using tile 'shelves'.

Horizontal chases can then be hacked at 200 to 300 mm centres
vertically. This operation needs to be planned with careful thought as
the chases should not be cut at or near a lift line which is a naturally
weak spot. The chases should not be cut in one operation as the
number required and the depth to which they need to be cut in relation
to the wall thickness will weaken the wall to such an extent that there is
risk of collapse. The ideal arrangement is to position the chases in the
centre of each course and to cut alternate ones. The chases can then be
brushed out, sprayed finely with water and filled with a stiff mix of
salvaged earth to which 10 per cent lime putty has been added to
increase strength. A plumb line set up over the eroded area allows the
depth of the erosion to be measured, together with the depth of the

chases. Plain clay roofing tiles must then be measured, cut and pushed into the groove to finish well short of the plumb line, with exuded earth packed back to ensure that the chase is well filled. Continue fixing the tiles with edges abutting to form solid 'shelves' and leave until firmly set before cutting the remainder of the alternate chases.

Once the rows of shelves are stable, the lowest eroded section can be hand packed with salvaged earth to which a little further clay, sand and fibre have been added and mixed with the minimum amount of water to produce a very stiff mixture. It needs to be tamped lightly but care needs to be taken to ensure that the upper tiles are not dislodged whilst the mixture is being packed at the top. Work can then continue, filling in each section until the top section has been completed and the mixture firmly tamped into the dovetail. Once completed, the entire eroded area should stand slightly proud of the wall face. The repair will take several weeks or even months to dry, the surface giving no indication as to the moisture content deeper in the wall. However, no attempt should be made to pare down the surface with a mattock until it is completely dry. Extra care should be taken during this operation to finish flush with the existing wall face as a strike to the tiles could break them and weaken the repair. The lift lines may then be restored with a small pick if required.

Cracks

Causes

Because of its low tensile strength, cracks pose one of the most common problems with earth buildings. The causes are many but can be categorized as being either a problem of the structure or of the site.

The clay content of the structure may be too high or unsuitable. In some parts of the country, the clay is highly expansive and cracks can be found in most buildings. In areas such as the New Forest, small cottages with up to eight full height cracks have been found and drastic measures have had to be taken to prevent their loss. Shrinkage cracks can be expected from any structure containing a high percentage of clay and water but few records exist of measurements taken as the building dries out.

One wall just over 5 m long built of rammed earth shrank 25 mm which is surprising considering the low initial moisture content. A recently constructed clay building built by the traditional method shrank between 20 and 25 mm over its short length of 4 m but a similar wall of chalk shrank only 14 mm over a 10 m length reflecting the different qualities of clay and chalk. The fibre content was usually sufficient to cope with the initial shrinkage and no further provision was made in the design of a building.

In Wessex, chalk boundary walls over 100 m in length were built without any form of movement joint and usually without any cracks. The few cracks which are found tend to be at corners or at the junctions between straight and curved walls where the tensile stresses are at their greatest because of a change in direction.

Cracks are sometimes found running between the top of the ground floor windows and the bottom of the first floor windows. Provided the openings have not been altered in any way, the rotting of the timber lintels over the ground floor windows is suspected as the cause of the problem. In those parts of the country where stone pillars were often incorporated around door openings, differential movement is caused by the rigidity of the stone set against the comparative flexibility of the earth resulting in cracks running from the top of the door opening to either the window above or the top of the wall.

The dead load of the upper floor and roof should be distributed evenly over all of the walls. An uneven distribution will encourage cracks to develop. The problem is worse in single storey buildings where there is no upper floor to counteract the load of the roof being taken on two opposite walls.

The quality of workmanship may also be responsible for cracks, particularly if the materials have not been mixed together thoroughly. However, a more likely cause may be poor construction of the under-pin course, allowing uneven settlement. Thermal movement would not normally cause cracking unless foreign materials with different coefficients of expansion have been introduced in a position where solar gain sets up uneven stresses.

A building should always be examined to check if alterations have been carried out. These put additional stresses on the fabric for which it was never designed and can result in the formation of cracks.

If the site of a building is made up ground, the degree of compaction will decide its load bearing ability, and any instability will give rise to uneven settlement resulting in cracks. A similar position arises when a firm base has not been reached before constructing the underpin course. On clay sites, the foundation trench needs to be very deep to combat ground movement and in chalk areas the building should rest on solid chalk rather than partly on subsoil and partly on chalk, regardless of the depth of excavation necessary, particularly if the ground is sloping. A trial hole will confirm the depth of the underpin course and ground conditions.

Ground subsidence due to mining operations is not likely to be encountered but nearby excavations can alter ground water levels to such an extent that the ground under a building becomes drier in some parts than in others, altering its loadbearing capacity. The movement of water through the ground is a common cause of cracking; leaking drains or water mains wash out fine soil particles leaving the larger ones to settle, causing subsidence.

Another common cause of cracks is the proximity of trees. When nineteenth century gardeners planted fruit trees against the earth walls surrounding their kitchen gardens, they were aware of the size they would achieve and set them an appropriate distance away. The neglect of walls following the demise of such gardens allows saplings to take root close to the underpin course, cracking the wall as they grow. No attempt should be made to remove such a tree and its roots, but it should be cut down to ground level, and its roots killed using a chemical agent which sets up a slow rotting process. It will take several years for the roots to rot, following which a certain amount of subsidence will take place. A regular inspection will alert the owner when action must be taken and the ground will need to be stabilized.

Trees planted a short distance from a building can still cause cracking if their roots extend to the underpin course, and break it up as they grow. As this process is unseen it is often overlooked, so nearby trees should be identified and their rooting systems ascertained to estimate the threat. If in doubt, a trial hole dug between the tree and the building will confirm suspicions about root length. Willows and poplars are the worst offenders. If the tree is still growing, it should be taken down and the roots killed. Any attempt to sever the roots near the underpin course will bring only short term relief before new shoots are formed which will inflict further damage.

If a tree is removed, clay in the subsoil expands as moisture moves in causing ground movement known as 'heave'. Its force is considerable and it will destabilize the underpin course and crack the wall. No action should be taken until the ground water level has returned to normal and the subsoil has stabilized.

Ivy is severely damaging to all earth structures (Figure 28). It requires nutrients and is not particular whether it obtains them from a horizontal or a vertical plane. Its roots will grow through the wall, appearing on the opposite side, and, as they expand, the wall will crack. Such cracks are usually hidden by the vegetation and it is necessary to kill the ivy before being able to examine the structure to see what damage has been inflicted. To achieve this, spray the ivy with a systemic weedkiller and leave it until the entire plant is completely dead, down to the tip of its roots and all its foliage has dropped. This will take several months and therefore must be planned well in advance. In chalk areas, a guarantee should be sought from the weedkiller manufacturer that the product will not react with the chalk and cause decay. Once the plant is dead

Figure 28 Ivy is severely detrimental to all earth walls. It has grown behind the render and already pushed off part of it. The remainder will follow soon. Having reached the coping, the tiles are dislodged to allow rain to enter the top of the wall. If not cleared away, falling render covers the underpin course allowing damp to rise into the wall and cause erosion.

(Figure 29), it can be severed, flush with the face of the wall, using secateurs but no effort should be made to remove those roots which are growing into the wall. Under no circumstances should attempts be made to pull the ivy away from the wall; collapse would then be inevitable.

The dangers of creepers growing on the face of an earth wall have been recognized to such an extent that tenants living in clay lump cottages on a private estate near Harling (Norfolk) in the late nineteenth century were expressly forbidden to allow them.

Tensile stresses build up in a wall which are released immediately it cracks. Sometimes, no further movement takes place and the crack can be repaired. Sometimes, however, further movement occurs which must be monitored before attempting repair. It is folly to repair a wall which is still moving.

Figure 29 Having been sprayed with a systemic weedkiller, the foliage dies, revealing a mass of dead roots which need to be severed flush with the face of the wall using a pair of secateurs. Great care is needed.

Simple measuring devices

Many types of measuring equipment have been devised, some of which are extremely complex, but the simplest type that is suitable for measuring cracks in earth walls comprises two large brass pins, each bent to form an 'L'-shape. Small holes are drilled either side of the crack and the shorter legs of the pins inserted and fixed with epoxy resin. The longer legs overlap each other across the crack and the end of each leg is marked on the other. Regular inspection will show whether movement is taking place and in which direction. On a rendered surface, a horizontal line can be scratched at the bottom of a crack, so that further movement will show as a continuation of the crack below the line.

If more than one crack appears, the wall between them becomes detached and is free to move in any direction. Each crack should therefore be monitored with at least two devices to enable the observer to decide which part of the wall is moving and in what direction.

Repairs

Before any attempt is made to repair a crack, all movement must have ceased, the cause should have been ascertained and the fault remedied. Figures 30, 31 and 32 illustrate various cracks, repair and damage to walls.

The most commonly seen 'repair' is where a cement-based mortar has been used to point the crack. This is not only unsightly but the mortar is denser than the wall and prevents it from breathing. The surface is also harder than the earth, aggravating erosion around the repair which is eventually left standing proud until it drops out.

Crack repair depends upon whether the wall is loadbearing or non-loadbearing and whether or not it is rendered. Buildings are always loadbearing and rendered if intended for human occupation but sometimes unrendered if used for animals, whereas boundary walls are largely non-loadbearing and often unrendered.

A fine crack with a maximum width of about 1 mm on a decorated rendered surface is eye catching and can cause concern whereas the same crack on an unrendered wall is invisible. Such cracks can

Figure 30 Subsidence has caused a crack which has been successfully held together with a purposely made iron strap, fashioned to the curve of the wall and bolted through with spreader plates on the back. Pugging the crack with chalk has prevented further erosion. The crack on the right is typical of those which occur at the junction of straight and curved work.

Figure 31 If left unrepaired, erosion takes place and the crack soon widens. Vegetation at the base grows through the crack, crumbling the wall as it grows.

normally be cut out and grouted in lime putty or grouted with the decorative material by brushing it well in during the regular maintenance routine.

A medium crack with a maximum width of approximately 5 mm in an unrendered boundary wall which is protected by a coping requires remedial treatment or else erosion will cause it to widen. The crack can be widened to the width of the scutch hammer and to a depth of at least 100 mm, brushed out, sprayed lightly with water and the salvaged earth mixed with a little clay, sand and water, hand packed and lightly tamped. Once dry, the repair is barely noticeable in a clay wall and almost invisible in a chalk wall. If the wall is rendered, the repair should be kept back from the face and a sample of the render sent for analysis to allow it to be matched when completing the repair. Both sides of the wall should be repaired in the same way, to avoid leaving a small, sealed air space in the centre.

Figure 32 Despite having been successfully underpinned, measuring devices revealed that the walls of this cottage were still moving. Massive erosion at the base had left insufficient earth to support the load of the upper floor and roof. The walls bulged and cracked, and only the emergency action to erect shoring prevented total collapse. The building was later demolished.

Alternatively, a medium crack can be grouted. This method of repair does not require any of the wall to be hacked away and only a thin line is left visible on the surface. Furthermore, the entire gap is filled in one operation rather than having to treat each side of the wall separately.

To grout a crack, spray finely with water and seal both sides of an unrendered wall with clay or putty or a rendered wall with a wide, strong adhesive tape but leaving a small gap at the top on one side. Mix one part of lime putty to three parts of fine sand and add sufficient water until it flows freely. The grout may then be poured in at the top of the crack using a small watering can and left to set before removing the temporary seals.

Larger cracks in loadbearing walls always need to be repaired as soon as movement has stopped.

The column method (Figure 33) avoids a straight joint right through the thickness of the wall but does not bond the two sides of the crack together. It allows slight movement to take place without destabilizing the repair.

Using a scutch hammer, cut a vertical chase either side of the crack to the centre of the wall, tapering the side to form a dovetail in cross-section and brush out the chase to remove the debris. Stabilize the salvaged earth by the addition of 10 per cent lime putty, mix well and add a little water to improve workability. Place a spadeful of the mixture in the bottom of the chase, tamp it well and continue to fill the

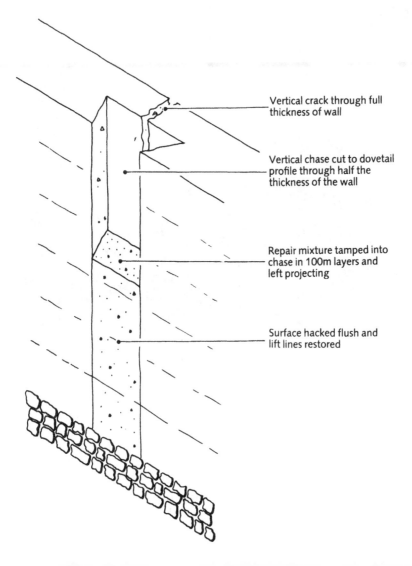

Vertical crack through full thickness of wall

Vertical chase cut to dovetail profile through half the thickness of the wall

Repair mixture tamped into chase in 100m layers and left projecting

Surface hacked flush and lift lines restored

Figure 33 Repairing a crack using the column method.

chase with the stiff mixture in thin layers, tamping each one until it has been filled and is standing proud of the surface. Once the repair has set, repeat the process to the other side of the wall, but spraying the back of the previous repair finely with water to allow the new work to bond. Once completely dry, the external surfaces can be hacked flush with a mattock and the lift lines reinstated to match the original.

The stitching method (Figure 34) attempts to tie the walls on either side of the crack together at regular intervals by the use of weak, *in situ* earth lintels with hooked ends. It is quick and simple to execute but makes no provision for any further movement following repair.

Using a scutch hammer, horizontal chases 200 mm to 300 mm high are cut into one side of the wall at about 800 mm centres, depending upon the length of the crack. Each chase need be no longer than 900 mm (i.e. 450 mm either side of the crack) and extend into the wall no further than one third of its thickness. Cut a square recess at each end of the chase to half the thickness of the wall and brush out debris before spraying finely with water. Stabilize the salvaged earth by the addition of 10 per cent lime putty, mix well and add a little water to provide a stiff but workable mixture. Hand pack the mixture into the

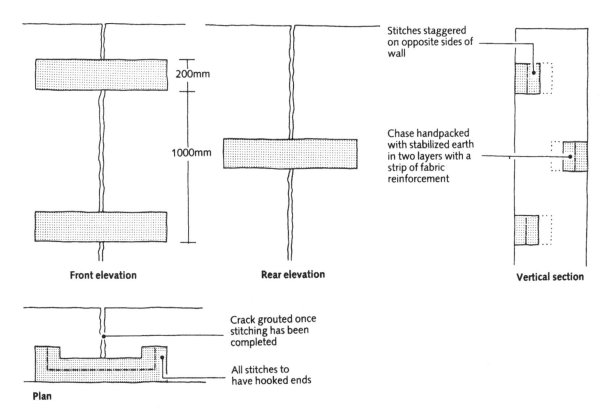

Front elevation   **Rear elevation**   **Vertical section**

200mm

1000mm

Stitches staggered on opposite sides of wall

Chase handpacked with stabilized earth in two layers with a strip of fabric reinforcement

Plan

Crack grouted once stitching has been completed

All stitches to have hooked ends

Figure 34 Repairing a crack using the stitching method.

chases to half the depth and tamp well. A strip of fabric reinforcement dipped in bitumen can be pressed into the damp repair mix and bent at the ends to allow it to be pushed to the back of the recesses. The rest of the chases can then be filled, tamped and left slightly proud of the surface.

The repair should be allowed to dry before the process is repeated on the other side of the wall, ensuring that each chase is cut between the previously cut chases rather than in line with them. When completely dry, the proud surfaces can be hacked back, the lift lines reinstated if desired and the crack grouted as previously described.

If a series of vertical cracks develops along the length of a wall, it may be desirable not to rebond them but simply to support the un-bonded areas. This can be achieved by constructing a buttress at each crack. The traditional brick buttress can add considerable character to a building but tall, slender buttresses may be constructed by building hollow precast concrete blocks on top of each other and threading them over steel reinforcing bars set vertically in the concrete foundation (Figure 35). Each block needs to be grouted in concrete as it is placed in position to secure the bars and packed in soft lime mortar

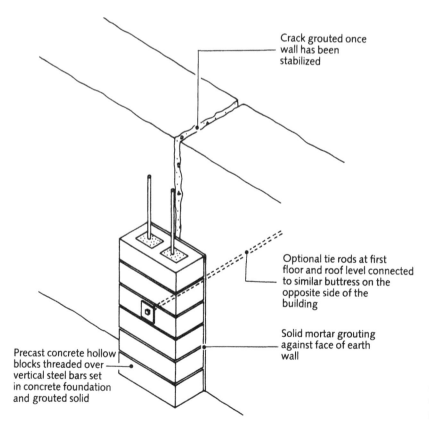

Crack grouted once wall has been stabilized

Optional tie rods at first floor and roof level connected to similar buttress on the opposite side of the building

Solid mortar grouting against face of earth wall

Precast concrete hollow blocks threaded over vertical steel bars set in concrete foundation and grouted solid

Figure 35 Supporting a cracked wall using precast concrete blocks.

against the face of the wall. The crack can then be grouted and the precast concrete blocks rendered to match the building.

Some buildings have been saved from collapse by constructing external concrete buttresses on opposite walls and connecting them with tie beams at first floor and roof levels. Such drastic measures should be considered as a last resort. They materially affect the character of a building and Listed Building Consent is required before execution.

Rat runs

The problem and deterrents

An inconspicuous hole usually about 70 mm diameter, just above the underpin course may appear to be nothing more than a minor problem but can easily be the only visible sign of a major problem.

The early builders were aware of the hidden danger of rat runs and developed deterrents, some of which may be encountered when probing the bottom course of a wall. The underpin course was sometimes built to an abnormally high level to counteract the problem, particularly in habitable buildings, a form of protection not always afforded to farm buildings. Where an underpin course does not exist, the threat of rodent infestation is even greater, although this may be offset by one or more of the many deterrents which are known.

In Wales, it is known that the bottom course was sometimes built of pure clay, to clog the paws of the scratching rodents. Other deterrents include a layer of gorse on top of the underpin course or the incorporation of nails, wire or broken glass in the bottom layer. Glass was sometimes powdered in the belief that it also deterred mice. Sometimes the bottom course has been built of stabilized earth to make burrowing more difficult. Even the mud and stud buildings of Lincolnshire were subject to this problem and the underpin courses were sometimes built off a plinth of compacted lime into which broken glass had been mixed in an attempt to deter the invaders. These anti-rodent precautions were taken because the early occupants were aware not only of the risk to health posed by Weil's Disease (spread by contact with rat urine) but also of the risk of the wall being weakened by burrows.

A well established colony of rats may live within the thickness of the wall and yet the occupier of the building may be quite unaware of their presence. The rat enters the wall from the outside, usually just above the underpin course, and burrows its way to the centre before turning along the wall and forming a tunnel around the building until it reaches a door opening. It will be realized that in cross-section, up to 20 per cent of the thickness of the wall will have been removed leaving only two outside pillars to take the load. On meeting an obstruction, such as a door frame, the rat will climb to the top and around the end of the lintel. On its journey, all fixing blocks cast into the wall are loosened causing the door frame to become loose and, eventually, to drop out if the door is slammed regularly. As the rodent family grows in size and more accommodation is required, tangential burrows are dug at different angles until the top of the wall is reached and the rats have access to the loft space where further nests are formed in the underside of the thatched roof. Eventually, the entire wall will become a mass of radiating burrows weakening the structure and causing cracking to such an extent that it is in danger of collapse (Figure 36). If this stage is reached, the erection of internal props and external raking shores becomes necessary to avoid the total loss of the building. It may result

Figure 36 Rat runs are difficult to locate, but a bad infestation can cause the walls to collapse. Only the erection of raking shores prevented total collapse to this pair of cottages whilst being converted into a single dwelling. The problem was exacerbated by roof spread and the removal of the party wall which had acted as a cross tie.

in a controlled demolition and rebuilding, if repair is no longer possible.

A rat hole should never be ignored. Debris at the base of the underpin course usually indicates that burrows are still being dug but if any doubt exists, the disappearance of baited food will confirm that the run is still in use.

Methods of location
Several methods of locating rat runs are available. The simplest is to use a long drill to bore small holes around the entry hole to determine the initial route taken. Further holes can then be drilled to trace the route, carefully marking the wall each time the drill hits the rat run and loss of resistance is felt. This method is time consuming and results in much abortive work, it is also unreliable as there is no guarantee that all of the burrows will be located.

A more reliable method is to blow warm air into the entry hole and to use an infra red thermometer, marking those parts of the wall where the heat is dissipated through the surface. Such thermometers are available on hire and consist of a gun capable of measuring temperatures from zero to 800 degrees Celsius and heat loss from 316 to 1999 watts per square metre. The wall should be marked out with a closely spaced grid and a reading taken at each intersection by holding the gun about 500 mm away from the face. Each reading should be marked and points of high energy loss joined up to show the line of the hidden burrows. This is a fairly slow method but the gun is cheap to hire and can be used by anyone without specialist training.

The most reliable method is to use a thermal imager. This comprises a battery operated video camera with a 12 degree lens and is capable of recording temperatures within the range of −20 degrees to +500 degrees Celsius with great accuracy. Such is its sensitivity that it is not necessary to blow warm air into the rat hole as the camera is able to detect the burrows by the difference in the air temperature of the tunnels compared with the cooler, solid core of the wall. By traversing the camera across the face of a wall, the air pockets are recorded on tape which can be played back on a normal video player. The cool, solid areas of wall show up as blue or green, whereas the warmer air channels are displayed in red or orange, allowing the rat runs to be clearly defined. The equipment is also available in a form which enables the image to be recorded on a compact disk for analysis by computer.

Once the burrows have been traced, they need to be marked on the wall surface using a grid system and confirmed by isolated drilling. Although a thermal image camera will produce accurate results, hire charges are expensive and a computer or video recorder is necessary to process the information. Furthermore, the task of transferring the information to the wall surface is time consuming and must still be checked for accuracy.

Repairs

Having located the runs and marked them on the face of the wall, the task of repair can commence once the rats have been killed. Search carefully for all of the entry/exit holes and place rat bait in each one every day until it is no longer being taken by the rodents. Precautionary measures need to be taken to ensure that other animals are not affected.

One method of repair is to seal the lower entry hole with damp, stabilized earth, well rammed in and drill a hole at the top of each rat run into which a liquid grout is poured and allowed to set. The grout is composed of one part hydraulic lime to three parts fine sand. Just sufficient water should be added to permit it to flow freely. If too much water is added, it will soak into the wall and weaken it even further. If the burrow pattern contains many horizontal runs, it will be necessary to inject the grout under slight pressure and to cut further access holes lower down the wall, stopping up each one before drilling at a higher level. The advantage of this method is that it is quickly executed with little or no visible signs of repair. However, the degree of compaction is unknown unless injected under pressure and the rats nests are left within the wall to rot and leave voids.

The use of hydraulic lime to stabilize earth should be restricted to those repairs within the core of the wall. It produces a fairly hard material, almost of cement quality, and should not therefore be used for surface repairs for fear of encouraging erosion at the interface of the hard and soft materials. The effect of adding hydraulic lime is to convert the earth into a weak mortar which will not dry in contact with air, like hydrated lime or lime putty, but which will dry and harden in contact with moisture by chemical reaction. The strength achieved is dependent upon the hydraulic capacity of the lime which is determined by the basic raw materials and the initial burning process. Once the lime has been added to the bedding mixture it should be placed in position within two or three hours. Any left over should be discarded.

An alternative method of repair is better suited to horizontal, low level burrows which are most commonly found. Starting from the bottom of the wall, small pockets are cut at regular intervals along the line of the rat runs using a drill and scutch hammer to allow access for the removal of debris and nesting material. The pockets can then be sprayed finely with water and handpacked with the salvaged earth stabilized with the addition of 10 per cent lime putty and mixed with a little water. The repairs should be well tamped and left standing slightly proud of the surface. Once dry, a second row of pockets are cut and the process repeated. A third and, perhaps, a fourth row of pockets may be needed as the work should be executed like underpinning and it is important not to remove too much of the wall at a time as this may induce collapse. Once completed, the entire rat run will have been cleaned out and repacked and the surface can be hacked flush with the wall using a mattock. This method is slow to execute and considerable care is needed to avoid stressing the wall unduly. For this reason, it may be necessary to relieve the walls of their load during repair by supporting the upper floor and roof on internal props.

Holes

Holes are usually caused by birds, rats, mice, wasps, etc. or by man who has cut a small hole through a wall for a pipe which has since been removed. If neglected, wind and rain will aggravate the hole, eroding its edges and enlarging it. This will be compounded if the hole is adjacent to a lift line which is a naturally weak point.

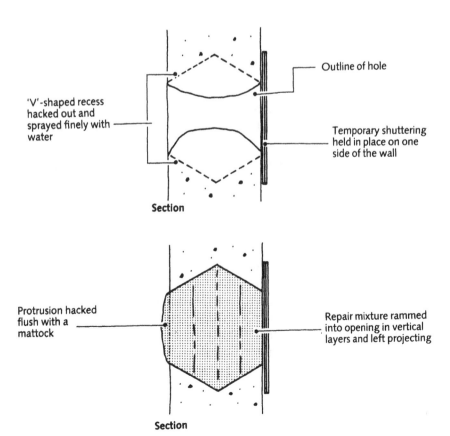

Figure 37 Filling in holes.

The most commonly seen 'repair' comprises several bricks wedged into the hole and set in mortar, usually with an eroded surface around it leaving the mortar standing proud. This is unsightly, inappropriate and provides only a short term repair whereas a permanent, sympathetic repair can be made using traditional materials with only a little more effort (Figure 37).

Using a scutch hammer, enlarge the interior of the hole to form a hexagon in both vertical and horizontal cross-section and retain the earth for analysis and reuse. Erect a strong temporary board on one side of the wall and support it with stakes driven into the ground – if at low level. At a higher level it can be hand held and supported by the weight of a man leaning against it. Brush out debris from the opening and spray finely with water. Having matched the clay, silt, sand and gravel from local sources, it can be mixed with the remainder of the salvaged material and a little water added if necessary to produce a damp mixture. The repair mix should be placed in the far side of the hole against the board and rammed in a vertical layer equal to about one quarter of the thickness of the wall. This should be followed immediately by three further similar layers until the opening has been filled. The repair is left standing slightly proud of the wall face. Once completely dry, a mattock can be used to pare back the protrusion and the lift line can be lightly hacked out with a pick if appropriate.

Adobe buildings

Although several of the previously described methods of repair are suitable for earth buildings of all methods of construction, except mud and stud, it is often more practical to repair an adobe building by cutting out damaged clay lumps, reconstituting the material into moulds and replacing them in the wall. This procedure has already been described as a method of repairing erosion at high level and the reader is referred back to page 94 *et seq*. It is best to cut out damaged areas to the full size of the block and remaining blocks may need temporary support to prevent them from breaking and falling. If smaller blocks are used, their size should tie in with the module, making due allowance for mortar joints.

The reason for the failure of the lumps should always be determined. A single mix was used to cast a number of blocks and any fault with it can be expected to manifest itself in the failure of other lumps in due course. The clay content of lumps is often lower than that found in *in situ* walling and varies between 5 and 15 per cent. The first to fail are usually those with a low clay content and these can be analysed to ascertain whether or not this is the cause of the problem, which can then be rectified when the salvaged material is recast.

When recasting, new fibre should be added and, if stabilization is considered necessary, an additional 10 per cent lime putty can be incorporated to ensure that the blocks are slightly stronger than the mortar.

Clay lump is a masonry type of construction and cracks may be stitched in the same way as brickwork. Cracks are rarely found and are usually due to external factors as opposed to faulty blocks. The blocks either side of a crack can be cut out and replaced using new mortar or recast and replaced in new mortar incorporating horizontal strips of stainless steel expanded metal built in at each level, if considered necessary.

Mud and stud buildings

Common problems

The last decade has seen increasing interest in the repair of these buildings and several have been given a new lease of life by enthusiastic volunteers working under the supervision of conservators, keen to acquire the skills which have been lost to the rural communities of Lincolnshire.

The main problem is the rotting of staves resulting in loss of the diaphragm. Daubed earth diaphragm walls are much thinner than other forms of earth walling and rely on staves for support. Once the staves rot to such an extent that they are no longer able to support the daub, it will fall, taking the entire diaphragm with it and leaving the interior of the building open to view through the lightweight timber framework. The strength of the staves is therefore of paramount importance together with the method used to fix them to the structural frame. This will depend upon the kind of tree from which the coppiced poles were cut, whether or not the bark was peeled off and to what extent the timber had been seasoned.

The later addition of an external skin of brickwork, which is commonly found is evidence of this problem. It is therefore surprising that mud and stud buildings were erected over such a long period without the technology advancing. A number of buildings are known to exist where the diaphragm walls have fallen, the building has been encased in brickwork and the inside face has been lathed and plastered, completely enveloping the oak framework and showing no sign of its existence.

Where daub samples have been analysed they show that a wide variety of materials have been used. A recent sample taken from Medlam, near Carrington contained almost pure clay with a little dung and straw, whereas samples taken further west have contained mud, lime, dung, road sweepings, ashes and even acorns.

Case study – Sotby

A sudden fall occurred at a cottage in Sotby in 1989. A warning that the staves were rotting had been given several years earlier when the daub pulled away from the diaphragm at low level and was 'repaired' by applying a strong concrete daub to the weakened staves. The decay continued and spread into the main framework causing the gabled flank wall to bulge and eventually collapse, suddenly, in an enormous pile of dust, leaving only the oak frame exposed.

The cottage is an excellent example of a mud and stud building and the initial advice from the local authority was to rebuild the wall in brickwork. However, specialist advice on the viability of using traditional techniques and materials was convincing and allowed to prevail. The rotted staves were discarded but the daub salvaged for reuse. Following repair of the main framework, with epoxy resin, new vertical staves of 37 × 50 mm softwood were creosoted and fixed to the oak with galvanized nails leaving 25 mm gaps.

Alternate layers of daub and straw, 75 mm and 25 mm respectively, were placed in a galvanized steel wash tub and sufficient water was added to mix it to a sloppy dough, with losses made up with a locally obtained Blue Kimmeridge type of clay. The daub was applied by hand from one side at a time, pressing hard to ensure that it filled the gaps between the staves. Three layers were applied externally and two layers internally finishing with a maximum thickness of approximately 375 mm at the base tapering to approximately 300 mm at the top. Once dry, the exterior was decorated with a mixture of lime putty and linseed oil and the interior with a mixture of emulsion paint and Polyfilla applied as a slurry, followed by a coat of neat emulsion paint.

The work was carried out by the owners, neighbours and volunteers, and the only problem encountered was the blending of the Blue Kimmeridge type of clay with the existing daub, where small soft spots have developed (Figure 38).

The use of locally grown saplings would have been better than sawn softwood for the staves but as the owner had a ready supply of sawn timber, they were used to enable an immediate start to be made on the repair work. It is possible that galvanized nails were used for the same reasons although stainless steel or non-ferrous nails would have been preferred.

The thickening of emulsion paint with Polyfilla is a practice unknown to the author and presumably was used to produce a textured finish. The use of a limewash or distemper might well have given the effect the owner required and would have been traditional.

Figure 38 Sotby: when the gable wall suddenly fell, the timber frame was repaired, new staves were fixed and the original mud reconstituted to build a wall 375 mm thick at the base. The work was executed by the owners of the building together with neighbours and other volunteers working under supervision (Lincolnshire Standard Group photograph).

Case study – Billinghay

The Old Vicarage at Billinghay was purchased by North Kesteven District Council and repaired in 1988 by a group of young people as part of an employment training scheme under the supervision of Peter Skaife. The work was administered by the District Planning authority. The building was stripped of its daub and staves to expose the timber framework on the front elevation but the outer skin of brickwork on the other three elevations was retained. The rotted timber framework was repaired or replaced, with new posts resting on metal shoes. The existing daub was salvaged and reconstituted, and losses made up with a peaty silt obtained from the River Skirth and straw to which 20 per cent cow manure was added. The daub was mixed in a large container in bulk and wheelbarrowed to the building, with a handful of hydrated lime added to each barrow load before being incorporated in the work.

New staves were cut from a local coppice and the bark was left intact to match the rotted staves just removed. They varied in size from 50 to 60 mm in diameter and were treated with Cuprinol before being used to form a new diaphragm. The staves were fixed vertically over the entire elevation leaving 75 mm gaps and secured to the frame with nails.

Shuttering was erected against the inside face of the diaphragm and a thick layer of daub applied to the outer face, pushed well into the gaps between the staves. This idea was abandoned as being impractical as problems were encountered in securing adhesion to the diaphragm. The shuttering was removed and the daub applied by hand from both sides of the diaphragm at the same time until a maximum thickness of approximately 225 mm was reached at the bottom of the wall tapering to approximately 175 mm at the top after shrinkage. The surfaces were finished on either side with a metal trowel to leave only the oak framework visible internally.

Once dry, the daub was protected externally with a brush-applied limewash, mixed on site from hydrated lime, to which tallow, obtained from a local abattoir, had been added in the ratio of 12:1.

It is doubtful whether the framework of any mud and stud building has ever been erected resting on metal shoes. The use of small stones is traditional and rot-proof and would have been preferred.

There is a mistaken belief that daubed walls contain much dung. Where it exists, the quantity is very small and the builders might have been better omitting it all together. The addition of such a tiny quantity of lime would have been difficult to disperse evenly throughout each barrow load of daub and its effect on the wall would have been practically nil. Lime putty is not easily available in mid-Lincolnshire but a limewash of better quality would have been obtained if quicklime had been slaked on site and matured, rather than using hydrated lime.

Case study – Bratoft
Whitegates Cottage on the Gunby Estate at Bratoft is thought to date from 1775 and is owned by the National Trust. It is a two storey building with mud and stud walls under a half-hipped and pantiled covered roof with a central flue. One end wall had been faced with brickwork and the daubed walls were rendered with a cement-based mortar. The cottage stood empty for 12 years until it was repaired in 1990.

The framework, of several species of hardwood, had rotted and been attacked by death-watch beetle but only minimum repair was necessary together with consolidation using epoxy resins. Two posts had to be replaced. The underpin course between the timber posts comprised three courses of brickwork on a foundation of compacted lime and broken glass and a bottle found in the foundations provided useful dating information.

The render was removed and discarded but the loose daub was taken from the external face to expose the staves in some areas before being analysed and retained for reuse. In other areas, the entire wall had to be stripped out to leave only the stuctural frame. The retained staves were treated with a timber preservative and losses made up with newly riven green ash obtained from a local coppice and split on site to match those existing, but left untreated. One row of staves, approximately 40 mm wide, were fixed with 30 mm gaps to the outer face of the frame at first floor level and one at ground floor level, using old cut nails to match the original.

Where the internal daub and staves were retained, one thick coat of daub was added externally. Elsewhere, one coat was applied to each side of the staves to give a finished overall thickness of approximately 200 to 225 mm for the full height of the wall. The salvaged daub was reconstituted and losses made up using clay excavated on the site and straw to which a little hydrated lime was added. It was both hand-applied and hand-finished smooth to match areas which had been retained.

Once dry, the exterior was protected by the application of four coats of Pozament polymeric limewash.

The clay pantiles were removed to expose a hidden covering of thatch which was renewed using combed wheat reed. The cottage is now inhabited but one small area has been left unrepaired for examination by visitors.

The reason for the addition of hydrated lime is uncertain. It would have increased the compressive strength of the daub but this is not relevant to a building whose weight is being taken on a lightweight structural timber frame. As a showpiece cottage, the use of Pozament is surprising. Although ideally suitable for external decoration, the opportunity of providing practical experience to the local workforce of making a traditional limewash has been lost.

Door and window openings

Cutting new openings

New openings should not be cut unless absolutely necessary. They weaken the wall and will damage it if not executed very carefully.

Design is important, the openings should cause the least impact to the wall. They should be as small as possible, as few as possible and spaced well apart. Their spans should be kept to a minimum and they should not be located near an existing opening leaving a column to take a point load. They should be kept well away from corners or other wall junctions for fear of breaking the bond at these vulnerable points and one door opening should never be formed over another, allowing the building to be split to its full height.

Lintels need to be inserted with considerable care to avoid cracking the wall. A scutch hammer or disc cutter should be used in preference to a hammer or bolster and two lintels inserted, one from each side of the wall, to support the full thickness. Having propped the floor or roof joists to relieve the load, hack out a horizontal chase to half the thickness of the wall. Its length should be equal to the width of the proposed opening plus a bearing of at least 300 mm on each side,

Figure 39 To provide daylight to the ground floor rooms of this barn conversion, the architect designed and installed Romanesque style windows based on a detail from the nearby farmhouse. Such windows eliminate the risk of cracking the wall, a risk which is always present when flat lintels are wedged and pinned to the underside of weak walls.

possibly more. The hardwood lintel can then be inserted and bedded in lime and sand before being very carefully wedged and pinned along its upper surface with slate slips. One tap too hard and the wall will crack because of its low tensile strength. It cannot be stressed too strongly that exceptional care must be taken over this operation. The entire process is then repeated on the other side of the wall.

It is now safe to remove the floor or roof supports. Two small vertical slots can be drilled through the wall directly below the lintel and a long cross-cut saw inserted with loose handles at each end. With one man either side the wall may be sawn down to the level required and the earth panel removed. A disc cutter may be used instead of a cross-cut saw working from either side before the centre of the wall is removed with a scutch hammer, but the saw produces a better, safer job.

The fixing of joinery can be a problem and is considered in more detail later. Door frames should be screwed to the underside of the lintel and the threshold plugged and screwed to the top of the under-pin course which may need to be grouted to provide a firm fixing. Window frames should also be screwed to the underside of the lintel and the sill screwed to dovetailed pressure treated battens cast into the top of the wall. For this reason, a window opening should be cut a little lower than the size of the frame to allow the top course to be rebuilt and the battens to be cast in. Provided that stout timbers are used for the joinery, jamb fixings are not necessary but a certain amount of lateral support is provided by the plastering and rendering of the reveals. Additional support is provided to internal door openings if a lining is used instead of a frame. Being the full width of the wall, its architraves can be fixed securely to wrap around the end of the walls on both sides of the opening to ensure that movement does not occur, even if the door is slammed.

All openings are points of weakness but the risk of cracking the wall when wedging and pinning the lintels can be overcome by forming openings with Gothic or Romanesque arches, preferably the former as they are stronger (Figure 39). A hole is drilled through the wall at the apex and a scutch hammer used to hack a vertical slot from the hole down to the required level which is widened on both sides to form the opening. The jambs are lined with brickwork built against the face of the earth wall and a centre erected to support the bricks over the arch until the keystone is inserted and the mortar has set. The gap between the brickwork and the earth wall, providing working space, is packed solid with salvaged earth to which a little lime and moisture has been added, and once the centre has been removed, a clear opening, lined with brickwork, is ready to receive the window or door frame. The brickwork can remain as a contrasting feature or be set back slightly from the face of the wall and rendered to match.

Enlarging existing openings
The daylight levels found inside a modern home cannot be matched by the small windows of an earth cottage. The windows were small because the low strength of the walls demanded that as much solid material as possible was provided to take the weight of the upper floor and roof. To reduce the solid:space ratio puts an additional stress on the wall and reduces its safety factor. Before enlarging the size of any opening, the matter should be considered very carefully and Listed Building Consent obtained if appropriate.

To increase the height of a window opening down to a lower level poses no problem, but the proportions of the building are upset and this may not be acceptable.

Increasing the width of an opening can be achieved provided that there is room to insert a new timber lintel over the existing lintel, otherwise there is no safe way of supporting the earth while the existing lintel is being replaced by the longer one. Two timber lintels must be provided side by side and inserted as previously described. Once the floor or roof props have been removed, the original, shorter lintel can be taken out and the opening cut to its new width using a long cross-cut saw. If the original opening is in the centre of the new opening, the original lintel will need to be removed, but if the new opening is alongside the original opening and incorporates it, the original lintel must be sawn to leave behind its bearing at one end.

If there is not room to insert a new timber lintel over the existing lintel, it may be possible to insert one underneath, provided that the existing lintel is one piece of timber for the full width of the wall. Having removed the door or window frame, the existing lintel can be propped and its bearings removed from one side of the wall and extended to allow a new, longer lintel to be inserted from one side. It should be supported on separate props, tightened so that it is in close contact with the underside of the existing lintel and wedged and pinned with slate slips on the underside at its bearing points. Both sets of props can then be removed and the process repeated the other side of the opening. The original lintel will remain in its original position supported by the new lintel.

Filling in openings and rebuilding fallen sections (Figure 40)
If rainwater penetrates the top of an earth wall, it is absorbed until it can hold no more. The clay coating to each particle of sand and gravel reaches its plastic limit and the wall will slump. The failure to maintain copings and to repoint ridge tiles is the usual cause in boundary walls

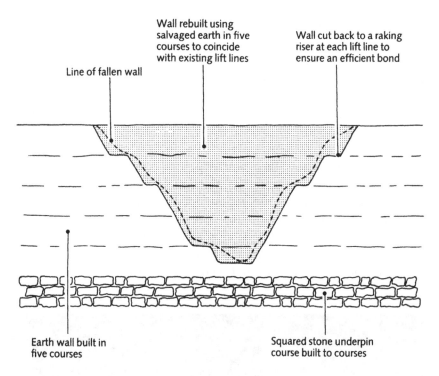

Wall rebuilt using
salvaged earth in five
courses to coincide
with existing lift lines

Wall cut back to a raking
riser at each lift line to
ensure an efficient bond

Line of fallen wall

Figure 40 Rebuilding a
fallen section of wall.

Earth wall built in
five courses

Squared stone underpin
course built to courses

but the opening left by the fallen section can be rebuilt using the
salvaged material, provided a few basic rules are followed.

Having attended to the coping of the adjacent sections to prevent
further loss, the fallen section should be tidied by controlled demolition. This will provide a firm base from where work can commence.
The height of this base will vary – only the minimum demolition
should be carried out. Ideally, it should be at its lowest point in the
centre, stepping upwards at each lift line until it reaches the top of the
wall revealing a stepped, 'V'-shaped opening and the riser of each step
raked to allow an efficient bond to be achieved when rebuilding. The
salvaged material should be analysed and reused, adding clay, fibre and
water and built up to its original height by the traditional method,
ensuring that each course matches the height of the original. This will
need to be supplemented with additional supplies of clay, silt, sand and
gravel gauged to match that existing. Once completed, the rebuilt
section will be of a slightly different colour and texture, and an algae
may form on the surface where spores have been absorbed by the fallen
soil and germinated in the damp atmosphere. This will die when the
moisture in the wall reduces to a level where the algae can no longer
live and any attempt to render it should be delayed until then. Once
rendered, all signs of the repair are hidden.

Large circular holes are often found in redundant agricultural buildings where cows have licked the surface in a regular place causing

erosion, and in extreme cases they can be large enough for the animal to pass through (Figure 41). If the holes reach almost to the top of the wall, little effort is necessary to cause the upper part to fall and the opening can be rebuilt as previously described. However, if the holes are at low level it is better to enlarge the interior of the wall to form an hexagonal cross-section, to erect shuttering one side of the wall and to ram in a damp mixture of earth to match the existing wall.

When doors and windows become redundant, it is better to lock them rather than remove them and fill in the openings. However, if it is necessary to fill them, several points should be borne in mind.

Care is needed when removing the door or window frame as they may be fixed to blocks cast into the wall and the wall will be damaged if they are dislodged. As the lintel will remain in place, the infilling wall will not be loadbearing but the jambs will be, so no attempt should be made to bond the two together. Earth may not be the most

Figure 41 Erosion caused by cows licking the surface of a wall can sometimes reach the stage where holes are formed. These can increase in size until the animals are able to break out of the cow shed. Only a high underpin course of brickwork has kept these cows within bounds. The roof is in obvious danger of collapse.

suitable material for filling in the opening and clay bricks are likely to be more satisfactory. They should be built in solid construction and lightly wedged and pinned at the top with slates to the underside of the lintel. Cavity construction is not recommended as condensation in the cavity is likely to soak into the earth. Concrete blocks should not be used as their co-ordinating size does not bond well with the wall thickness. They are heavy, more difficult to handle than bricks and do not adapt to changes in temperature so well without causing movement. No attempt should be made to plaster the infilling wall until the mortar is completely dry. A lime plaster can then be used which will cause the minimum of cracking at the junction with the existing work.

A more traditional approach would be to provide a framework of timber studs on both sides of the opening and to finish it with sawn or riven timber laths and three coats of lime plaster. The end studs can probably be fixed to the existing fixing blocks but intermediate studs need to be supported by the lintel and a new sole plate framed to the end studs.

Fixing to earth walls

Unless provision was made when the wall was built, it is difficult to achieve a strong fixing to an earth wall, and it is better avoided altogether if possible. Buildings of stabilized earth, rammed construction or adobe are stronger than those built by the traditional method and suitable nails may be used, but fixings cannot approach the strength of those in masonry walls and alternatives should be sought.

Wall plugs are of little use unless they are very long. They need to penetrate the wall at least 150 mm to be effective and screws are no longer made to this length. Screws should always be used in preference to nails as they are more controllable during insertion and are less likely to crack the wall.

The builders were aware of the problem and allowed for it in the design. Dovetailed shaped timber fixing blocks were cast into cottage walls at underpin course level to allow skirtings to be fixed. They were cast into the jambs of openings to allow door and window frames to be fixed and into boundary walls to allow hooks to be screwed in for securing tying wires upon which climbing plants were trained.

Unless provision was made during construction, fixings are more difficult but the problem can usually be overcome. The material is weak in tension, particularly at the lift lines which should always be avoided. Long stainless steel 'Gunnebo' nails with a thick, barbed shank can be obtained which increase resistance to extraction by having a larger surface area in contact with the earth. As they are gently tapped into the wall, they twist slightly but should be fixed with care

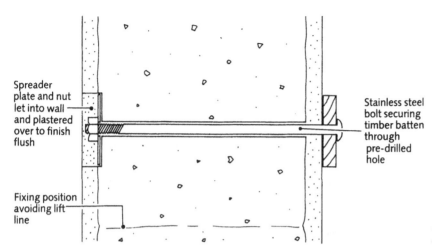

Spreader plate and nut let into wall and plastered over to finish flush

Stainless steel bolt securing timber batten through pre-drilled hole

Fixing position avoiding lift line

Figure 42 Fixing to an earth wall.

for fear of breaking the bond. Resin bonded anchors are likely to produce a loose plug of solidified earth.

The possibility should be examined of pilot drilling through an earth wall, gently tapping in a square section of sawn hardwood block, providing support to the back face, pilot drilling the block and screwing into the end grain. Alternatively, drill through the wall, insert a stainless steel rod with threaded ends, fix a spreader plate each side and use nuts to tighten them. This holds the wall in slight compression and leaves the ends of the rod projecting to which fixings can be made. One side can be let into the wall and plastered over if necessary (Figure 42). A line of rods can share a common timber batten to which a variety of items may be fixed. Rainwater downpipes can be attached in this way using long coach screws with distance pieces to ensure that an overflowing pipe will not damage the wall (Figure 43). A building known to the author has been completely clad externally with clay tiles hung on battens attached to a series of posts which have been bolted right through the walls to spreader plates fixed internally.

Fixing points can be provided in the top of a wall by removing part of the top course and rebuilding it using the salvaged material with a little lime putty added, and casting in sawn fixing blocks of dovetail section. This method is useful for fixing window sills but should not be relied on where heavy pressure is expected.

The quality of clay lumps varies in East Anglia and it is known that the rules of an estate near Harling forbade any tenant to nail into the walls for fear of damage. Despite this, some buildings in the region were faced with brickwork which was tied back to the clay lump with metal ties which were nailed. Recent tests have shown that a pull exceeding 0.17 kN is necessary to dislodge the strongest of them.

Figure 43 Rainwater downpipes are best kept well clear of an earth building to avoid erosion. Firm fixings should be at gutter and underpin course level and stabilized at swan-neck level with long stainless steel bolts fixed right through the wall.

When converting or altering an earth building, it is better to dispense with fixings as far as possible. Skirtings were not always fitted and may not be necessary, but, if essential, they could be formed *in situ* of hydraulic lime and sand. Heavy items such as kitchen wall cabinets are better supported on a brick wall built alongside the earth wall and stabilized by steel straps fixed to the underside of the roof or upper floor. Alternatively, if headroom necessitates a space over the units, the brick wall is better built full height to provide firm fixings.

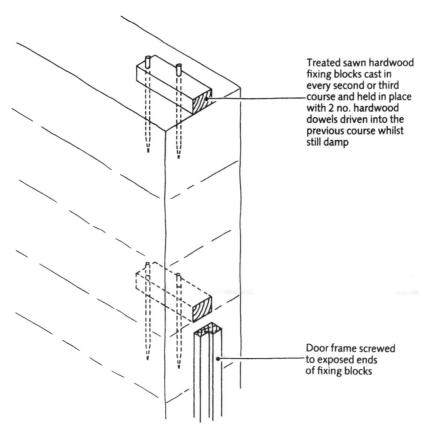

Treated sawn hardwood fixing blocks cast in every second or third course and held in place with 2 no. hardwood dowels driven into the previous course whilst still damp

Door frame screwed to exposed ends of fixing blocks

Figure 44 Provision for fixing joinery.

When a new building is being constructed, provision can be made for fixing joinery (Figure 44). Mr Alfred Howard of Devon casts in treated sawn softwood blocks to window and door jambs as the courses are raised. Each block has two holes through which pointed dowel rods are gently hammered to hold the fixing blocks in position and one edge is left exposed to which joinery can be screwed. The same principle is used to secure wall plates at first floor and roof level and the best results are obtained when the previous course is still damp.

Installation of upper floors

The need to increase accommodation within the parameter of an existing building by the installation of an upper storey has already been mentioned, but various matters should be considered before a decision is made. The redistribution of the weight of the structure on the walls and the ground is the main consideration and a structural engineer should be employed to calculate the consequences of such a decision.

If only part of the building is having an upper floor inserted, the implication is more acute.

Upper floors are usually inserted in buildings such as barns which were intended to be single storey structures. The walls may not be capable of taking the additional load and to make them do so is to encourage problems. The ground is usually capable of taking the extra weight which is small in comparison to the total weight. The weight of the wall is much greater than the weight of the upper floor and roof combined and is in the region of 5 tonnes per lineal metre for a wall 500 mm thick. It is safer to assume that the existing walls are incapable of taking the additional load safely and to support them on an independent structure. This leaves the existing structure unaffected by the new floor provided that no attempt is made to tie the new and the existing structures together.

Should the weight of the floor be taken on a clay brick skin built inside the existing earth wall, a cavity of varying thickness can be maintained to eliminate damp, improve thermal efficiency and provide a sound base for the wall plaster. It is not wise to fill the cavity with gravel as this reduces air circulation and the gravel may become clogged with fallen earth causing the cavity to be bridged. A damp proof course should be provided under the new wall together with a stepped cavity tray at the third course to reduce this problem.

Neither is it wise to seal the cavity at the top, and the provision of outlets will ventilate it and reduce the risk of condensation. It may be wise to continue the inner skin at first floor level and to ventilate the cavity into the ventilated roof space, but care should be taken not to transfer any weight of the roof structure from the earth wall to the brick wall as the existing walls rely upon it for lateral stability.

On the debit side, the space taken by the walls will result in smaller rooms which may need careful detailing around the openings to avoid damp penetration. The loss of character to the inside of the building is considerable and may destroy the charm which provided the initial attraction. There is also the possibility that the inner face of the earth wall may deteriorate without any obvious sign until it is too late to repair it, unless it has first been rendered.

Should the weight of the floor be taken on timber beams supported on columns and piers the disadvantages associated with an inner skin are eliminated. However, point loads are created on the subsoil which may require foundations of considerable size and depth to ensure stability. Columns can be avoided if brick piers are constructed in the corners of the rooms adjacent to the existing earth walls and separated from them by a vertical damp proof course. Once plastered and decorated to match the walls, they are unobtrusive and may even appear to be part of the existing structure. The edge of the floor boarding must be scribed to the profile of the earth wall. It is not usually practical to cover the gap with a skirting fixed to the wall but a foamed polyurethane gap filler can be used which is masked by the carpet.

Installation of heating systems

Most schemes to alter, convert or extend an earth building allow for the installation of a central heating system. No earth structure was ever designed with this in mind, but it presents no problem provided that adequate precautions are taken to protect the fabric of the building.

A wet heating system involves the installation of radiators and connecting pipework through which water is pumped from a heat source which may be fired by gas, oil or solid fuel. Problems may arise with the fixing of the radiators, and floor mounted models are preferred. However, where wall mounted models are essential, the brackets should be fixed to vertical timber bearers, secured to stainless steel threaded rods, fixed through the wall with spreader plates on the back, as previously described. Radiant and convected heat from the radiators should be prevented from drying out the wall locally by providing a reflective insulating fabric behind each one. This will also improve the efficiency of the radiators. Conducted heat from the pipework should be isolated where it passes through the walls for the same reason. Oversized pipe sleeves are essential with an insulator filled around the pipe and which can be sealed with a cover plate each side of the wall for neatness.

The location of the heat source should be chosen with care. All fuels produce gases which need to be expelled into the open air and the construction of a flue and chimney is the traditional choice, although the introduction of balanced flues allow a considerable saving in costs. A balanced flue requires a large hole to be cut through the wall which can be made by drilling a series of closely spaced holes around the perimeter and removing the centre section with a scutch hammer. It should be cut oversize, sleeved and insulated with mineral wool to prevent local drying out of the earth. Purposely made cover plates can be fitted on both sides with provision made for sealing the joint externally.

The construction of a brick flue and chimney is preferred, and the fireplace is usually made a feature of the building. The outlet of the boiler discharges directly into the flue where it rises and dispels into the open air without the necessity of cutting a hole through the earth wall and fitting an unsightly balanced flue. The stability of a structure is enhanced if the flue is situated centrally but an existing earth wall should not be cut vertically to provide for its insertion as the cross tie will be lost and the outer walls may start to bow. An independent structure is therefore better but it may be located adjacent to an existing wall provided that it is not bonded to it in any way and that the two are isolated by a layer of insulation. A substantial foundation is necessary to cope with the high point load. The flue should be lined to cope with the temperatures likely to be encountered depending upon the choice of fuel to prevent damage to the mortar joints caused by the gases.

The installation of a warm air central heating system is not so detrimental to the building, as the need to fix radiators to walls and to circulate water through pipework is eliminated. The same principle rules apply and care must be taken in the location and cutting of holes for ducting, their insulation from the internal walls, and the extract of flue gases. A fireplace, flue and chimney are not necessary, as free-standing insulated flues are obtainable with roof terminals but their appearance may not be acceptable.

The installation of electric storage heaters is the most appropriate of all heating systems in an earth building. Wiring can be threaded through conduits and fixed in the floor or roof space with connections to fittings let into wall chases and plastered over. The heaters, although large, are floor mounted and therefore present no fixing problems. However, when placed near earth walls, they should be insulated. As a flue and chimney are not needed, they are ideally suited to barn conversions where obstructions to the roofline are not acceptable.

Earth buildings have thick walls and the inner surface is heated by the installation of a central heating system. Intermittent use will prevent the structure from warming and the risk of condensation and mould growth will arise. A low, steady heat should therefore be maintained throughout the heating season.

Installation of damp courses and membranes

Those used to living in a modern dwelling expect the inner skin of the external cavity wall to be dry. When those same people move to a rural area and purchase an old building with solid earth walls, they expect the same standards but their attempts to achieve it can cause damage to the building.

Damp should not be prevented from rising through the underpin course into the earth wall. It should be encouraged to evaporate on the outer surface before it reaches the wall so that a physical barrier is not necessary. The method of achieving this has already been described. If the underpin course is high, the damp will not rise into the earth wall.

The original builders made no attempt to inhibit the passage of moisture although various materials were available to them had they wished to do so. To insert a barrier will dry out the wall to such an extent that its loadbearing capacity is considerably reduced. The safety factor of earth walling, although adequate, is much lower than it is for masonry construction and to reduce it further will bring it nearer to the threshold below which the structure may become unstable. Compression tests recently undertaken on samples taken from a wall during reconstruction have shown that the earth was 20 per cent weaker when dry than when it had a moisture content of 6 per cent.

It is therefore advisable not to insert a damp proof course which will restrict the natural flow of the moisture, prevent the wall from breathing properly and reduce its loadbearing capacity. Due to these problems a number of major installation companies will not carry out work to earth buildings.

In some parts of the country, underpin courses tend to be built higher, particularly in later buildings. Examples can be found where they reach the entire height of the ground floor of a two storey building. The chances of damp rising above this level into the wall are remote and the installation of a damp proof course at ground floor level treats the underpin course rather than the wall and has no effect on it.

Should damp rise to an unacceptably high level above the underpin course, the excess moisture will put the stability of the structure at a greater risk than if it is very dry. The reason should be sought and the fault remedied. The usual causes are broken drains running parallel with the outside wall of a house or a pressurized water main leaking at the point where it passes through the underpin course.

The application of damp proof membranes to the upper surface of the ground floor slab of a building can divert damp to rise into the adjacent walls. Most agricultural buildings were built without floors but cottages were usually built with floors of compacted clay or chalk, sometimes stabilized with skimmed milk and capable of breathing. In the last century many were paved with brick or tiles, bedded in lime stabilized earth, but still able to breathe through the joints. To seal such a floor with modern waterproofing products inhibits the natural process of evaporation causing the moisture to travel horizontally under the membrane until it reaches the underpin course where capillary action pulls it up into the wall to an unacceptably high level. The problem is not common but buildings at greatest risk are those in low lying areas with seasonally high water table levels. It can be remedied if the water table level is reduced by installing land drains around the site, but this is a drastic measure which could have unforeseen effects on other parts of the building. Inside the building, the provision of waterproofed building paper well lapped and laid loosely will allow the floor to continue breathing and provide a sound base for laying carpets.

If, on the purchase of a property, it is discovered that a damp proof course has been installed, one should be aware of the problems which might arise. However, no action should be taken to remove the damp course unless such problems are manifest. Excessive drying and dusting of the wall surface, the appearance of fine cracks for which no other cause can be found and the lack of adhesion between the wall and the render causing it to fall, are all signs that the wall is too dry. It is not possible to remove a chemical injection type of damp barrier but the electro-osmotic and porous tube types can be removed if absolutely necessary. It is better not to interfere with these barriers as such action can only stress the building, but it may be wise to experiment by rendering the outside face of the underpin course, providing a bridge which will allow a little moisture back into the wall and enable the dusting surface to stabilize itself and cracks to seal themselves. There is no chance that the render will rebond itself to the wall but the clay coated sand and gravel particles will adhere to each other in much the same way as they did before.

Buildings of rammed earth are of later construction. Their walls are denser and stronger and were built during the nineteenth century when it became standard practice to incorporate a damp proof course on top of the underpin course. The development of the railways enabled slate to be distributed throughout the country and this was the material usually chosen to keep rising damp away from the earth structure. The slates were laid to the full thickness of the wall at a point where the render ceases. For this reason, the damp proof course is not usually visible and its ability to function is in doubt, as it may have been cracked when the first layer of soil was rammed into the shuttering.

Boundary wall coping details

The necessity to protect the top of a boundary wall coupled with the difficulty of fixing to earth walls has stretched the imagination of the builders to produce a variety of details to support the coping. Some details are common to a single village, some are found scattered throughout a region and some are found only in isolation. Whether or not these are contemporary with the wall is in doubt as old photographs show that when thatched copings required renewal, many were replaced with other materials.

Walls were built around produce gardens of farms and country houses to provide security for the crops, to protect them from cold winds and to provide a framework on which fan trained fruit could be grown. Openings were few, even large gardens only had one opening wide enough to allow a cart to pass through. The coping to the wall also acted as a deterrent to any would be scavenger who attempted to scale it. He would discover that tiles loosely bedded in earth fell off with such ease that attention was immediately drawn to his violation of the garden.

There are four coping materials usually encountered, thatch, the most traditional, tiles, slates and corrugated steel sheet.

Thatch

Details vary little and usually involve a series of timber bearers set at regular intervals across the thickness of the wall, overlapping generously on both sides and held in place by a triangle of solidified earth comprising fine chalk, or clay and sand, mixed and laid by the thatcher. The difference in texture between the solidified earth and the earth wall is clearly seen in the products of the craftsmen of different trades. An eaves soffit was provided of timber slats fixed on top of the bearers or with hazel spars interwoven between the bearers. The coping of longstraw or water reed thatch was then fixed to the triangle of solidified earth and held in place with hazel spars. It was sometimes finished with decorative stitching. The ridge was sealed in various ways according to local custom.

Tiles

Tiled copings are those most commonly found today with variations depending upon the degree of pitch, the thickness of the wall and whether plain or pantiles have been used. The simplest detail was used for thin walls – a triangle of solidified earth on top of the wall into which a single course of nibbed pantiles were pressed whilst the earth was still moist. Half-round copings were fixed on the ridge and bedded in lime mortar. Where the thickness of the wall demanded two or more pantiles each side, the bottom edge was supported on a longitudinal batten held in place with iron angle stays screwed to the ends of fixing battens cast into the wall at regular centres during its construction. A similar principle was used to secure several courses of plain nibbed tiles with a double course at the eaves to prevent rainwater running into the top of the wall. A wide eaves soffit was provided of similar tiles laid side by side and held in place by the solidified earth. In some cases, several courses were built corbelled to create an extra wide soffit. Where a thick wall had a coping of several courses of plain tiles, better quality work demanded that they be fixed with pegs to a groundwork of longitudinal battens supported on a framework of triangular timbers held in place vertically with solidified earth.

Slates

Slates provide a lightweight alternative to tiles, and fixing details are similar but as only one row of slates either side of the ridge is usual, they need to be overlapped to prevent rain from entering the wall. The most commonly found detail comprises slates centre nailed to a wide, longitudinal batten supported on a framework of triangular timbers set upright and held in place with solidified earth. A second batten is usually fixed at higher level to support the top of the slates where the ridge tiles are bedded. An alternative detail consists of timber bearers set at regular intervals across the thickness of the wall, overlapping either sides and held in place by a triangle of solidified earth. The ends were splayed to the pitch of the triangle and a longitudinal batten nailed on top to support the bottom of the slates. A ridge timber was let into the top of the solidified earth which supported the top of the slates. The slates were overlapped at the edges and fixed both to the ridge and the batten before clay ridge tiles were bedded in mortar. An elaborate variation is known where the ridge board is laid flat and supported on short vertical struts built into the top of the wall at close centres. The slates were butt jointed and the joints covered with narrow strips of slate held in place at the top by the ridge tiles bedded in mortar.

Corrugated steel sheet

The introduction of curved sheets of corrugated steel enabled protection to be provided quickly and cheaply, though to the detriment of the village scene. The most simple detail comprises timber battens cast across the thickness of the wall at regular intervals to which two wall plates were nailed. The steel sheets were laid across the wall with the minimum of eaves overhang, drilled on top of the corrugations, nailed to the plates at close centres and coated with tar to prevent rusting. Sometimes the timber grounds cast into the wall to support horizontal wires were used to fix vertical iron straps which held the wall plates in position, dispensing with the need for cross bearers.

Missing tiles and rotted timbers need urgent replacement if a boundary wall is to survive. Temporary protection in the form of tarpaulins held down with ropes and bricks, will prevent further decay. The groundwork should not be removed without first making a sketch of the detail which should be retained and used when the new coping is constructed to ensure continuation of the local tradition.

Stabilizing walls

Introduction

In those parts of the country with suitable soils, the craft of earth walling became highly developed and the products of the early builders remain to delight the modern visitor. In other parts of the country, where clays are more expansive and less cohesive, and where a well graded range of sands and gravels is not available locally, the earth wall builder struggled to provide a durable building. Their lack of permanence has led to their demise in recent years leaving only the best examples which may require major repair if they are to survive.

In less well constructed buildings, faults develop which may be impossible to repair and to attempt to do so may jeopardize the stability of the building. In such circumstances it may be appropriate to stabilize the building, that is to secure it in its present state and make it safe, preventing faults from deteriorating further.

The building should be examined to determine two main points. It must be established whether or not the structure is still moving. Diagnostic devices may be installed at strategic points and the measured information collated to determine the buildings stability or direction of movement. Remedial work cannot be undertaken on a structure which is still moving. An unstable building is dangerous and could collapse without warning. The erection of temporary external shoring and internal props is therefore an essential first step to be taken to allow the second point to be determined in safety: the reason for the movement then needs to be established so that it can be rectified. It may be due to the unsuitability of the natural materials from which the walls were built, roof spread because of inadequate design, the result of later alterations, or changes in site conditions which have affected the load-bearing capacity of the ground.

Having ascertained the stability of the building and the reason for its instability, the extent of the repair work can be determined. The intention should be to stabilize it permanently in its present position, to prevent any further movement and to prevent collapse. This can be achieved in various ways depending on how the instability manifests itself.

Falling corners

A corner is a naturally weak spot and loss of bond at this point is a common problem. Once this occurs, each adjacent wall is free to move in any way it pleases. In the past, such walls were usually stabilized either by the introduction of tie rods or by purposely made wrought iron corner straps.

To insert a tie rod involved drilling a hole through an unstable wall, threading an iron rod through it and anchoring it to a stable part of the structure. Circular end plates or decorative cross straps were then fixed on the end of the rod, on the outside face of the wall, to spread the force over as large an area as possible. Sometimes the rod went right through the building and spreader plates were fixed on both outer faces. Walls with a high tensile strength could be partially corrected and pulled back into their original position by applying heat to the centre of the rod allowing it to expand in length. The plates were then tightened and the walls pulled together as the iron cooled and contracted. However, it is doubtful if this procedure was ever attempted on an earth building.

Tie rods are of dubious value to earth buildings and are better suited to masonry construction, which is stronger in tension. They will not prevent further movement of the structure, as the spreader plates simply pull through the wall by shear action. In addition, they place a strain on the anchorage point, which is often the opposite wall, causing further damage. If anchored to a timber upper floor or roof structure, the diaphragm of the boarding is utilized to improve the stability but in an unboarded roof, they can impose excessive lateral forces on the joists which the bearing points are unable to absorb, resulting in loss of bond at an already weak point. Examples may be found where tie rods appear to have been successful but, on examination, the plates are usually loose and the rods slack. In such circumstances, their introduction has been of no benefit to the building which had obviously stabilized itself and they can normally be removed with safety. The introduction of tie rods is not recommended.

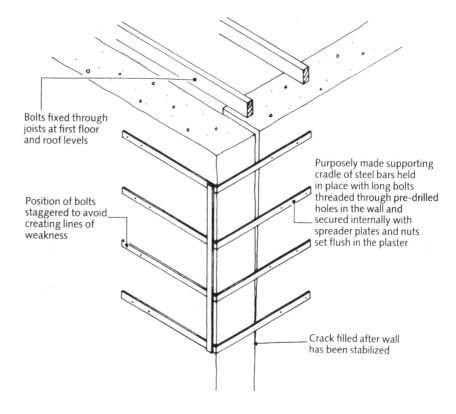

Bolts fixed through joists at first floor and roof levels

Position of bolts staggered to avoid creating lines of weakness

Purposely made supporting cradle of steel bars held in place with long bolts threaded through pre-drilled holes in the wall and secured internally with spreader plates and nuts set flush in the plaster

Crack filled after wall has been stabilized

Figure 45 Stabilization of a falling corner using steel straps.

Falling corners were sometimes stabilized with a series of straps (Figure 45). Based on a different principle to tie rods the straps relied upon the two adjacent walls either side of the crack for their support, whereas tie rods relied on an opposite wall or timber beam for support. In either case, once stabilized, the cracks could be filled.

Several long, flat iron bars were bent at right angles and fixed horizontally around the corner of the building at regular centres for the full height of the crack. They were secured to the two adjacent walls

Figure 46 A successfully repaired falling corner. Wrought iron straps were bolted through the wall to spreader plates on the inner face and the cracks were filled. The work may appear unsightly but early action prevented the loss of the building. The work was executed many years ago.

with several long iron bolts which went right through the wall and were fixed on the inside face with nuts and small spreader plates. Where the straps were adjacent to first floor and roof timbers, longer bolts were used to connect to them, tying the corner back to the structural diaphragm. The angled straps were connected together on the outer face by a vertical flat iron bar with bolts. The vertical bar was not always fixed at the corner; if fixed to the ends of the angle straps it was likely to be bolted through the wall as well.

Figure 47 The conversion of the attic space into habitable accommodation has increased the loading to beyond the limit of the wall. Alterations to the first floor window opening and an overflowing rainwater head may have contributed to the cause of the falling corner. The owner chose to demolish the entire building and replace it with a new one built of brickwork rather than rebuild the fallen section.

The installation of corner straps was a bold action and their degree of success is unknown, but their existence has ensured the survival of a number of buildings which would otherwise have collapsed had such remedial work not have been executed (Figure 46). They are interesting but ugly in appearance and Listed Building Consent would most probably need to be obtained if their installation was considered today. The installation of corner straps should only be considered where a crack is in the early stages of formation and should not be relied

Figure 48 Three storey buildings are rare. The thickness to height ratio of the walls of this chalk house was extremely high and both upper floors sagged due to undersized joists. Joinery pulled away from the walls, doors would not close and windows were pushed out of alignment. Both flank walls bulged and the corners started to fall. The building has now been demolished.

upon to support major structural failure (Figures 47 and 48). Where installed, they should be non-ferrous to avoid rusting.

The use of cementitious grouted anchors may be a viable alternative to prevent small corner cracks from getting bigger where it is important that there are no visible signs of support (Figure 49). Their installation at the first sign of a corner crack may prevent collapse and rebuilding at a later date.

Each anchor comprises a stainless steel tube which is capped at one end but has a hole in the side of the tube near the cap. The tube is encased in a sleeve of muslin type material which is sealed at both ends. Holes of a diameter slightly larger than the tube are drilled through the corner of the wall on skew, adjacent to the crack at 300 mm centres vertically and the tubes inserted to allow a cementitious liquid grout of cement and very fine sand to be injected into the ends of the tubes under slight pressure. The grout is forced through the hole near the capped end to expand and fill the muslin sleeve, holding it there by friction and preventing the two parts of the severed wall from moving further apart. The grout will set, retaining the bond, and the only visible sign of the anchor is the end of the tube through which the grout was injected. This can be made good with mortar and decorated to effect an invisible repair.

Cementitious grouted anchors are manufactured in different diameters from 10 mm upwards, to suit a wide range of repair conditions. They are available in lengths of up to 10 m and have been designed for use with concrete and masonry structures. Their use with earth walls is limited because of the low tensile strength of the material and they should not be relied upon to stitch together major structural cracks. They can also be inserted at T-junctions where the bond between the external and internal wall has been broken.

Tie rods, corner straps and cementitious grouted anchors are all methods of stabilization which cause tensile stresses within the structure because they do not cure the source of the cracking. If the cause is

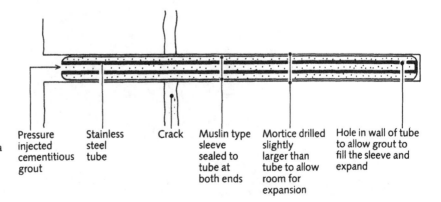

Figure 49 Stabilization of a falling corner using cementitious grouted anchors.

| Pressure injected cementitious grout | Stainless steel tube | Crack | Muslin type sleeve sealed to tube at both ends | Mortice drilled slightly larger than tube to allow room for expansion | Hole in wall of tube to allow grout to fill the sleeve and expand |

thought to be ground settlement, consideration should be given to underpinning the structure or stabilizing the ground to increase its load-bearing capacity. If the cause is thought to be due to roof spread, then the principle behind the design of the roof should be considered and checked by a structural engineer before introducing remedial timbers.

Buttresses need not be unsightly and may improve the appearance of certain types of earth buildings, particularly if they are single storey (Figure 50). If properly designed, they will prevent any further

Figure 50 Buttresses need not be unsightly. Lack of internal walls to this thatched chapel left only the roof trusses to act as tie beams. As they were not adequate, the walls bulged but were restrained by five brick buttresses built each side of the building in line with the roof trusses. An earlier attempt to stabilize the walls with a bund wall of brickwork had failed.

movement of a falling corner and may be constructed in conjunction with the installation of cementitious grouted anchors.

The construction of buttresses affects the character of a building and Listed Building Consent will most probably have to be obtained before their erection (Figure 51). It is of prime importance that they are well designed, not just from an aesthetic point of view but also from a practical point of view. If not, their construction can lead to more problems than they solve because of the additional weight added to the side of the building which is already leaning. A solid foundation taken down to firm subsoil is essential to prevent any chance of the buttress subsiding, causing the building to lean further. As this foundation is excavated next to the face of an unstable wall, great care is needed and it may be necessary to erect temporary shoring, to be removed once the buttress has been completed. Aesthetically, it is more appropriate to construct the

Figure 51 The clay walls of this two storey cottage had eight full height cracks and each section started to move independently. They were stabilized by vertical buttresses built of hollow precast concrete blocks reinforced with steel rods and grouted. Listed Building Consent was obtained as the buttresses affected the character of the building.

buttress in brickwork than any other material but the use of precast concrete hollow blocks threaded over steel reinforcing bars and anchored to the roof structure have been used occasionally (Figure 51). On no account should attempts be made to bond a buttress into an earth wall. It should simply abut and be bedded in a soft lime mortar. The top should be weathered to slope away from the wall to prevent rainwater from saturating the structure. If a rendered and decorated finish is desired the brick joints should be left recessed to provide a key for the render to be applied later once the mortar has completely dried.

Occasionally, corner cracks have been ignored resulting in the emergency erection of raking shores and internal props to prevent collapse. When this has occurred, a carefully controlled partial demolition has been necessary involving the eventual removal of the shores but retention of the props. The missing section must then be rebuilt to take the weight of the structure before the props can be removed. Demolition and rebuilding should be considered the very last resort, but all effort should be directed towards saving as much of the historic structure as possible.

Figure 52 The flank wall to this chalk terrace bulged, and was supported by raking shores for several years. The owner chose to carry out a controlled demolition and to rebuild it using flint lime bricks, dense aggregate concrete blocks and aerated concrete blocks. Different coefficients of expansion have resulted in a series of cracks appearing in the render which require regular grouting.

It is not always practical to rebuild in earth, as problems arise regarding bonding to the existing walls. However, this may be possible depending on circumstances and advice for filling in openings is given elsewhere in this chapter. The rebuilding may have to be undertaken in modern masonry cavity construction off a new concrete foundation. The use of clay bricks is recommended as the amount of thermal movement is minimum as opposed to sand lime bricks or concrete blockwork where considerable movement takes place resulting in a permanent fracture between the new and existing walls (Figure 52). On no account should an attempt be made to bond the brickwork to the earth wall but stitches, as previously described to repair cracks (page 107), can be introduced at these two vulnerable points to ensure the stability of the existing structure. The brickwork can be lime rendered on completion. It is not easy to match the finish of the original although the application of several coats of limewash will help both to soften the impact and to grout the fractures between the new work and the existing building.

Leaning walls
Walls which lean can be alarming. Inevitably, they lean outwards and the cause needs to be identified positively and remedial action taken before the wall is stabilized, although the erection of raking shores is always a wise precaution as temporary protection.

To decide whether or not it is safe, measurements need to be taken on site and a section through the wall drawn to determine the centre of gravity (Figure 53). A vertical line should be drawn from this point to bisect the base of the wall. Divide the thickness of the wall at its base into five equal parts. If the vertical bisects the outer two parts, then the wall is suspect and temporary supports are advisable. This rule of thumb guide tends to err more on the safer side than similar guides given for masonry structures which are bonded together with either cement mortars, producing a higher tensile strength, or with lime mortars which are more flexible. When a wall develops a lean, one side of the sheer line is in compression and the other in tension. This is manifest in the formation of horizontal cracks on the tension side and spalling on the compression side, accompanied by falling debris.

The cause of the lean is due either to action by weather, design fault or alterations to the building or several of these reasons.

Subsoil is affected by weather conditions which can reduce its loadbearing capability. If the outside soil is wet and the inside soil is dry, movement may occur under the foundation causing the outer part to tilt downwards, taking the wall with it. As this problem usually occurs under walls with a wide concrete foundation, it is not likely to

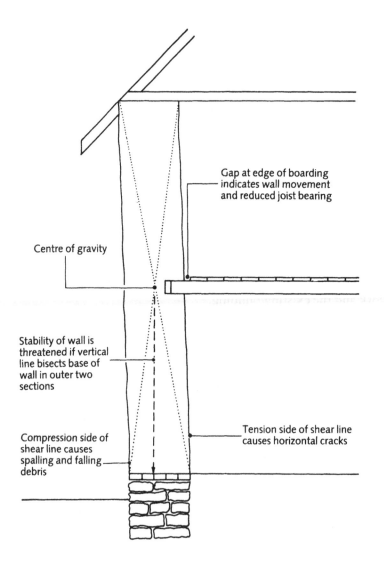

Gap at edge of boarding indicates wall movement and reduced joist bearing

Centre of gravity

Stability of wall is threatened if vertical line bisects base of wall in outer two sections

Compression side of shear line causes spalling and falling debris

Tension side of shear line causes horizontal cracks

Figure 53 Ascertaining the stability of a leaning wall.

be encountered under an earth wall unless it is extremely thick. Where soil is higher one side of a boundary wall than the other, it acts as a retaining wall, a role for which it was never intended. The movement of groundwater is inhibited causing a reservoir to form on the side of the wall on which the soil is higher. This will tilt the wall if it freezes. If the subsoil is high in clay, excessively wet weather will cause it to expand and hot, dry weather will make it contract. This seasonal movement causes cracks to develop in the underpin course allowing in debris which clogs the cracks and prevents them from moving back into position when the moisture content of the ground is reduced.

A wall may lean because of a fault in its design and its thickness to height ratio should be established to decide whether it was built too high in the first place. It is difficult to give 'safe' ratios as this varies from one locality to another depending on the basic materials found and the method of construction. However, a rule of thumb guide can be ascertained by the measurement of other structures in the area and making a comparison. The thrust from a badly designed roof will also cause its supporting walls to lean and both sides of the building should be checked. A lack of ties and collars allows the roof structure to drop and spread, thus pushing out the walls. The threat of further movement can be minimized by the introduction of new timbers to stabilize the roof.

Alterations made to a building can easily cause its walls to lean. Overloading, the removal of internal cross tying walls and the demolition of adjacent structures, allowing dry soil to become saturated for the first time in many years, are all potential sources of trouble. Recent ground disturbance due to the installation of services can cause problems, particularly if not well compacted when backfilled. Such disturbance will also affect the movement of ground water but this would probably only be a short term problem. The way in which the building is used should not be overlooked. Increased loading due to first floor living or the storage of heavy items in the roof space can affect the stability of the walls and cause them to lean.

A leaning wall can affect a structure in several ways and remedial works may be necessary to elements other than the wall. The wall should first be checked to determine whether it was built with a battered face as was once customary in some areas. The end bearings on lintels should be examined for signs of movement and consolidation if necessary. The joists to the upper floor and roof need to be examined to determine the length of bearing still active. Gaps alongside the edges of floorboards at first floor level can usually be measured to assess this point. The bond with an internal cross wall will have been broken and a vertical crack opened up. This breaks up the cellular plan on which many buildings rely for neighbourly support but the cracks should not be grouted until the leaning wall has been stabilized.

It is important to determine whether or not the wall is still moving. This can be ascertained by fixing a permanent bracket on the face of the wall, suspending a plumb line from it and carefully marking its position on a permanent plate let into the ground. By re-suspending the line from the same bracket at monthly intervals, the marks on the ground plate will show the amount of movement.

To stabilize leaning walls, various methods have been employed, determined largely by the cause of the problem. The use of tie rods and buttresses have already been considered for stabilizing fallen corners and both have been used to counteract leaning walls. However, their degree of success is dependent on accurate diagnosis of the fault and work has often been executed which has served no useful purpose. The use of cementitous grouted anchors may be considered to rebond the external wall to the internal cross wall if the lean is slight, but this method should not be relied upon to stabilize a major lean. A leaning wall is weaker than a vertical wall and repair and stabilization work should be kept to a minimum to avoid reducing its strength even further. The installation of tie beams and spreader plates would serve little useful purpose, and a row of them would perforate the wall creating a line of weakness at its weakest point. If the cause is due to low loadbearing capacity of the soil, underpinning should be considered as the first option if the lean is not too great. Otherwise, well designed buttresses at strategically located centres determined by a structural engineer are likely to be considered the most practical form of stabilization.

It is possible to tilt masonry walls back to their original position but due to the fragililty and low tensile strength of earth structures, this procedure is not recommended, even if a cradle is made to support the wall during the tilting operation.

Subsoil problems are the most likely cause of leaning walls and they cannot be stabilized unless the loadbearing capacity of the soil is increased. This can be achieved either by underpinning or by ground stabilization.

Underpinning involves increasing the depth of a wall to reach a new, deeper, firm foundation. A number of earth buildings have been underpinned but it is an operation which needs to be designed and executed with considerable care if the structure is not to be put at risk. The underpin course is invariably only the same width as the wall and there is no solid, *in situ* concrete foundation to support the wall while it is being underpinned. Underpin courses of small stones such as flints were sometimes constructed by filling them into a trench and pouring in a runny mortar of sand and lime to fill in the interstices and to hold the flints together. To excavate beneath the flints causes them to drop and it is difficult to underpin such a wall. It is not always safe to underpin courses of larger stones unless trial holes excavated at regular intervals along the length of a leaning wall show them to be well bonded together.

Perhaps the most suitable type of foundation to underpin is one built of bricks which usually incorporated footings up to twice the thickness of the wall. The bricks are bonded together and are capable of supporting each other while the operation is carried out, provided that it is executed in very short lengths with no more than one quarter of the wall being underpinned at any one time. The entire work will therefore take four separate operations to complete it. For more detailed information on underpinning, the reader is referred to *Foundation Design and Construction* by M. J. Tomlinson.

One method of increasing the loadbearing capacity of the subsoil is ground stabilization. Any local soft spots can probably be excavated and backfilled with concrete. However, if the problem is along the entire length of a wall, a specialist engineering company should be asked to consider the possibility of stabilizing the ground. This can be achieved either by grouting or by micropiles.

Specialist grouting is used in the civil engineering industry in a wide range of applications. Grouts may be based on greases, bitumens, resins or modified cements to solve different problems. Cement-based grouts are particularly suited to soil stabilization and may incorporate pulverized fuel ash (PFA) from power stations, high temperature insulation (HTI) powder based on fireclay from furnace linings, and specially imported fine sands as binders. These ensure high strength, resistance to sulphates, rapid drying and deep penetration. The effect of the grouting is to increase the loadbearing capacity of the ground to prevent any further movement of heavy structures.

The 'Pali Radice' patented system of micropiles is available as an alternative method of ensuring that the subsoil is capable of supporting an increased load. The system involves the installation of a grid of cast *in situ* bored piles, which can be as small as 75 mm diameter. They are inserted at regular centres along both sides of the wall through the underpin course and into the ground at an angle of 60° to depths of up to 30 m, crossing each other like scissors. The piles are bored with rotary drilling rigs to eliminate vibration and flushed out with water before a cement and sand grout is pumped in under pressure.

Once filled, steel reinforcing rods or small cages are injected and the grout is left to set. The installation equipment is capable of boring through any obstruction with the minimum of noise and is portable enough to enable it to be operated in areas of restricted access and headroom. The effect of the system is to transfer the weight of the structure from the subsoil to the micropiles.

Bulging walls

A bulging wall is one which curves outwards either on plan or in vertical section or, more probably, both. A wall may lean and bulge, and methods of stabilization therefore need to be cross-referenced to achieve a suitable repair.

No confusion should exist between a bulge and a slump. The former appeared after the building had been erected whereas the latter occurred during the construction of the building and was usually corrected immediately by paring down or rebuilding. A bulging wall shows small cracks on the face in tension but does not normally show any sign of distress on the face in compression.

To monitor movement, permanent plates and temporary plumb lines can be fixed as previously mentioned (Figure 54). An approximate indication of the scale of the problem can be gained by holding a long length of straight timber vertically against the face of the wall, checking it with a spirit level and measuring its distance from the base of the wall. When drawn in section, the thickness of the wall at its base should be divided into five equal parts and if a vertical line drawn from the centre of gravity bisects the outer two sections, remedial action is recommended and temporary support should be provided.

The most usual cause of bulging walls is alterations made to the building, but the thickness to height ratio should be ascertained and compared with other local stable structures to eliminate this design fault. The increased loading of the upper floor due, perhaps, to higher dead loads such as the creation of a further bedroom can cause a vertical

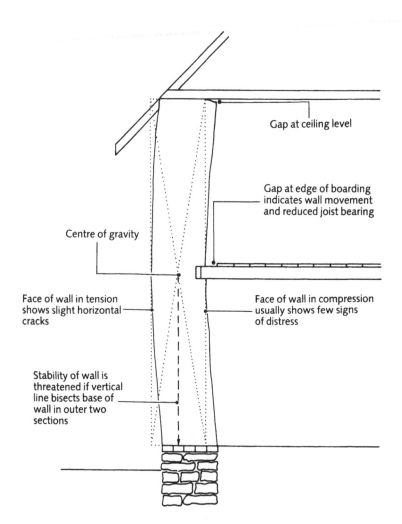

Gap at ceiling level

Gap at edge of boarding indicates wall movement and reduced joist bearing

Centre of gravity

Face of wall in tension shows slight horizontal cracks

Face of wall in compression usually shows few signs of distress

Stability of wall is threatened if vertical line bisects base of wall in outer two sections

Figure 54 Ascertaining the stability of a bulging wall.

bulge which can be measured at the centre where a gap will be found between the edge of the floorboards and the face of the wall. The increased loading may be due to the way the building is used, such as first floor living, resulting in higher imposed loads which also create a vertical bulge. The loss of bond between the internal cross walls and the outer wall will allow a horizontal bulge to develop, and if the wall is restrained by heavy ties at roof level, a vertical bulge will also develop. The same problem will occur if timber tie beams are removed from agricultural buildings (Figure 55). In those parts of the country where it was usual to support the upper floor joists on wall plates, any sign of rot will render them incapable of taking the load placed upon them, allowing the joists to drop and causing a vertical bulge. The removal of adjacent structures which once offered lateral support is another possibility which should not be overlooked and enquiries should always be made to eliminate this cause.

Figure 55 The owner of this long, narrow potter's drying shed cut the low level timber tie beams because they interfered with his use of the building, but the ends were left embedded in the external walls. Since then, cracks have appeared at the corners and the walls have started to lean outwards. Only short, high level collars couple the rafters.

The effect of a bulging wall is similar to that of a leaning wall except that the subsoil is obviously strong enough to take the load placed upon it and ground conditions are not therefore affected.

Similar methods of stabilization are also suitable involving cementitious grouted anchors and buttresses. The installation of two tie bars bolted through the wall to the first floor joists or beam with a connecting, spreader strap may be appropriate if the bulge is only slight, but the strap would need to be let in flush with the outside face of the wall to achieve an efficient bearing.

Figure 56 A brick wall built against a leaning earth wall will not prevent collapse unless built very strongly. Shoring was necessary to stabilize the end wall of this cottage despite a one brick thick skin having been built to enclose the entire building. The front elevation also bulged from the pressure of the moving mass of earth.

In the past, one method of 'repairing' a bulging wall was to build a brick wall in front of it, and to render and decorate it to match the rest of the building (Figure 56). To avoid reconstruction of the roof at eaves level, the wall was kept as thin as possible resulting in a half-brick thick skin which was sometimes connected with tie rods and spreader plates to a similar skin constructed on the opposite side of the building. Such measures did not prevent further movement of the earth wall and the bulging was often transferred to the brickwork even if built one brick thick.

Externally, these buildings show little sign of having been constructed of earth as the end walls were often also clad to tie the front and back brick skins together.

Drastic measures such as these are not recommended. Neither are elaborate schemes to cast *in situ* concrete ring beams or to chop out vertical chases and cast in reinforced concrete columns and connect them with tie rods. Such methods of stabilization are irreversible and involve too much damage to an already weakened structure. They represent overpowering ideas more suited to masonry construction and put the stability of the entire structure at risk.

Bulging walls can be a major problem and if there is still movement, they can be dangerous. The last resort, both in the interests of safety and to save the building from collapse is a controlled demolition and rebuilding. A total demolition is never necessary but those parts of the structure which are at risk should be identified and only the minimum amount of demolition should be undertaken.

Consolidation

Research work is being undertaken in several countries into consolidation to reduce the rate of decay of earth buildings, but little is thought to have been commissioned in the United Kingdom. This involves treating a structure with a highly penetrating liquid which will improve the efficiency of the binder and thereby ensuring its stability.

Whereas dampness is the main enemy of earth structures in temperate climates, wind erosion is the principle threat in drier climates and most research has been directed towards the latter.

Most consolidants tend to inhibit the porosity of a wall and are not therefore recommended for use in temperate climates. Waterproofing liquids, although not completely waterproof, also prevent free moisture movement and can do more harm than good. If friable surfaces are encountered, they can be stabilized with a 10 per cent ethylpolysilicate solution in water, sprayed on finely. However, the formation of a thin skin can easily part company with its host exposing a new vulnerable surface.

Patented bonding agents are available from builders merchants which can be diluted and used as a sealer of porous friable surfaces. Being PVA emulsions complying with BSS 5270 (1989), they are water permeable but are not recommended for consolidating the surface of earth buildings as they are intended for internal use on dry backgrounds.

To be effective, deep penetration is necessary. In theory, walls of pure chalk should be capable of being stabilized using limewater but several finely sprayed coats would need to be applied daily for many days to have any effect. Limewater will not inhibit the flow of moisture but the degree of saturation necessary to be effective could put the stability of the structure at risk. It should be used with caution.

Long term trials being undertaken at Grenoble (France) and Fort Seldon (New Mexico) include consolidation by chemicals such as isocyanates, acrylics and silanes applied by impregnation or spray. Results are awaited. Research carried out in Australia with water repellents has resulted in the moisture absorption rate being reduced to 1 per cent of that of an untreated wall and similar research in India has produced a dry repellent in concentrated form which is mixed with water on site and applied by spray. It can also be used as a damp proof course in new work. Trials are made on the basis that decay can be reduced if resistance to rain and wind can be increased.

The problems posed by earth building in the climate of the British Isles are often based on damage caused by salt nutrients deposited in a wall by damp rising through capillary action. No attempt should be made to consolidate such walls as all forms of pre-blocking treatment will only hasten decay.

Fire damage

Earth is an incombustible material but its chemical composition can be altered by heat. Chalk structures are at a higher risk than clay but the degree of change is decided by the intensity of the heat.

An open fire burns at an average temperature of 200° C but can vary between 50° C and 300° C. A closed fire however burns at twice these temperatures and if forced ventilation is introduced, the temperature will rise considerably. The amount of heat generated is dependent on the fuel: group 1 house coal melting (i.e. forming clinker) at 1425° C, group 2 at 1200° C and group 3 at 1040° C.

A house fire is not likely to reach these temperatures but if chalk reaches 880° C it drives off the carbon dioxide and water content isolating the calcium carbonate and changing it into calcium oxide (i.e. quicklime). The use of water to extinguish the flames will turn the building into a slaking pit with all its associated dangers.

To drench an earth structure in water will put out the fire but can cause considerable damage to the walls leading to total collapse. The risk is greatest in very cold weather when the water soaks into the walls, freezes and exfoliates during the thawing process. Modern fire fighting equipment allows water to be extracted at almost the same rate as it is applied, but saturation will still occur and the structure will be at risk until it has dried out. Temporary protection must be given to keep frosts away from the wet areas and this is easily achieved by stacking bales of straw either side of the wall and clearing them away when all risk of frost has passed in late spring. The warmer weather will dry out the structure very slowly and this may take up to two years. Until then, the building should be considered unstable and no rush made to reinstate it.

Demolition

Demolition is the last option and should be a forerunner to rebuilding, if repair is considered impractical. The material maybe reused but will

Figure 57 Demolition work was abandoned on this house of puddled clay when spot listing was considered. To rebuild this rear wall was a practical proposition but the building was eventually demolished and the site cleared for new development.

need fibre adding to it, as the original fibre will have rotted and provision must be made for controlling shrinkage. It is wise to analyse the material to ascertain the binder:aggregate ratio as it is probable that additional clay should be added to make up losses caused by the demolition (Figure 57).

The task of demolition should not be underestimated. A chalk wall is reconstituted rock and if constructed by the rammed method, is as difficult to demolish as chalk rock is to excavate by hand. A terrace of chalk cottages in Kings Somborne (Hampshire) was demolished in 1962 to clear the site for a new bungalow, but a tractor fitted with a front loader was unable to achieve the task even after constant battering. A steel rope was therefore placed around the buildings and attached to the tractor in an attempt to pull them down: it cut through the buildings like a cheese wire, leaving them standing.

A complete boundary wall may be bulldozed, but if only part is to be taken down it must be executed with care. It is necessary to isolate it from the rest of the wall and this is best achieved by sawing into vertical sections and pushing them over, to be broken up later with a

Figure 58 A controlled demolition in progress. Only the front wall of this cottage was to be demolished so the upper floor and roof were propped, but supporting and access scaffolding needed to be provided for the safety of the workmen in the event of a sudden collapse. Work was much slower than had been anticipated.

sledge hammer. If a wychert wall is bulldozed, it breaks up into large crystalline blocks weighing over one tonne and requires further demolition on the ground.

If only part of a building is to be taken down, a controlled demolition should be carried out (Figure 58). This is a slow task that involves isolating the area to be rebuilt from the sound structure by sawing, and breaking it up using a hammer and bolster. The more sound the walls, the more difficult the task and a controlled demolition recently monitored by the author resulted in two men taking one hour to demolish one square metre of hard chalk wall 500 mm thick. The workmen expressed surprise at the strength of the wall, echoing earlier recorded comments of a demolition gang taking down a rammed earth building in Surrey during the Second World War.

Once an earth wall becomes saturated with water, it will slump. However, this is not recommended as a method of demolition due to lack of control. Considerable time is needed to allow the structure to absorb sufficient water and there is no warning of its collapse. Once that occurs, the site becomes an uncontrollable quagmire making operations impossible. The wet earth must be moved clear of the structure and is not suitable for rebuilding until it has dried out.

If it is proposed to demolish an extension to an earth building, whether it is built in earth or masonry, it is wise to inspect the structure that will remain after the demolition has been completed. The extension may have been acting as a buttress to a weak structure and its removal may cause the collapse of that part of the building which is to be retained. Considerable remedial work may then be necessary to support the structure and prevent movement.

Chapter Six

Wall protection

EXISTING METHODS OF PROTECTION

It is usual to offer some form of protection to the face of a wall, particularly if it is part of a building. Such protection improves the durability of a wall and combats erosion and damage caused by frost and animals. Any form of protection offered should not harm the wall in any way and must ensure that the porous, breathing capabilities of the structure are maintained.

Some forms of protection are inherent in the design, such as adequate eaves overhang and a high underpin course, with special consideration being given to buildings where additional protection is necessary. Farm buildings were sometimes built with rounded corners around which cattle manoeuvred regularly, thus eliminating vulnerable points and strengthening the wall. Stables are found with brick or stone door jambs to eliminate wear caused by abrasion as the horses entered and left the building daily.

The most commonly found form of protection is the external render. To be successful, it must not be stronger than the wall as earth buildings have a weak surface and can only accept a weak render. Walls of rammed earth and adobe are stronger but it is not wise to use modern renders which are better suited to concrete blocks or clay bricks.

Analyses of samples taken from different parts of the country have shown that many locally derived renders have been developed with varying degrees of success. Many of these are based on lime and sand and often reinforced with animal hair, but additives such as brick or coal dust were sometimes employed to give a distinctive appearance or salt added to ensure that it retained its porosity. At Houghton Lodge (Hampshire) the render to the wall surrounding the kitchen garden contains 7 per cent lime, 9 per cent hair and vegetable matter, 32 per cent crushed chalk and 52 per cent sand, measured by volume, whereas at nearby Ramridge Park brick dust and additional lime replaced the chalk on a render of similar date.

These two examples show how renders vary in an area where the most commonly found form of protection often consisted of nothing more than a slurry of chalk and water. In Ireland, the renders were sometimes of lime and sand but more usually they copied the structure and comprised tempered clay, fibre (hair or fine straw) lime and dung. In some parts of mainland Britain, lime putty and cow dung have been used, mixed in equal proportions and in Northamptonshire, it is known that neat cow dung was applied to walls in the winter, only to be removed in the spring and sold as garden fertilizer. In other areas

Figure 59 Expanded metal lathing should not be fixed to earth walls prior to rendering. The weight of the render pulls the lathing away from the wall due to the difficulty of obtaining a firm fixing.

mud renders of clay and sand were frequently found and in East Anglia it is common to find agricultural buildings protected with tar applied either directly to the face of the clay lumps or to a render, sometimes with a sand blinding thrown on. Renders were applied directly to the face of a wall, to a natural key cut into the earth, without any need to resort to any form of mechanical key.

The widespread availability of Ordinary Portland Cement together with the introduction of plastering contractors to repair or replace renders has been responsible for much damage to and loss of earth buildings throughout the country. The days when an owner would gather natural, locally found materials to repair his cottage and out-buildings have gone. The rise in the use of plastering contractors, together with the introduction of various forms of mechanical keys in an attempt to complete the work more quickly have been detrimental.

The most commonly found defect of the present century is a hard, thick, heavy, cement-based, impervious render sometimes with a pebbledashed finish, adhering to a rusty mesh of expanded steel fixed to the face of the wall with long, hammered in nails. The fixings are not adequate and can damage the wall beyond repair as the

Figure 60 The entire render has fallen off in one sheet and brought the face of the wall with it up to 100 mm thick in places. Galvanized nails 150 mm long had been used to fix the metal lathing to the chalk wall. A traditional render of lime putty and sand applied directly to the face of the wall would have been much better.

heavy render pulls away the mesh (Figure 59). Unfortunately, such 'protection' is still being applied and will provide problems for the years ahead.

Figure 61 Two years after the first visible sign that the render had lost contact with the wall, it has started to collapse revealing a rusty steel ribbed lathing. Condensation has taken place behind this thick, hard, heavy render causing erosion to the wall 100 mm deep. Even the underpin course has been rendered.

Where a mesh key is provided, the render will adhere to it allowing gaps between the back of the render and the face of the wall. If the wall contains lime and dung, the two can react in the unventilated atmosphere to allow anaerobic bacteria to develop and weaken the wall. Even if lime and dung are not present, deterioration can still take place due to condensation forming at the back of the render causing the outer surface of the wall to expand and lose the few bonds which exist between the render and the wall (Figure 60). One need only tap the outer surface of a render and listen to the hollow sound to determine the extent of the problem.

Cement-based renders are sometimes applied without the use of a mechanical key, with equally disastrous results. It is important to allow moisture free movement within the wall and this is often absorbed from the ground through the joints of the underpin course by capillary action. Such moisture is free to travel to the surface of the wall, and evaporate, but if a hard render has been applied, the movement is inhibited and the moisture will travel higher up the wall seeking an outlet. Rising damp up to 3 metres is known. The accumulation of pressure within the wall is released once the render cracks and is forced away from the face allowing rain to enter, driven by strong winds and exacerbated by cold, frosty weather. The wall is eroded behind the render and the debris falls and builds up against the underpin course (Figure 61). In severe cases large areas of render will fall, some of which can weigh more than a man.

It is not advisable to remove a cement-based external render and to replace it with a more suitable one unless it has completely failed. To do so would cause parts of the wall to fall away with the render in those areas where the adhesion is the greatest.

PATCH REPAIRS TO RENDERING

When repairing an old, original render, the aim should be to replace the minimum that is necessary so as to preserve as much of the original as possible. A sample of the render weighing at least 100 grams may be sent to a laboratory for analysis and report for use as a guide to matching the original. However, analysis will only identify the constituent materials and may be misleading, as aggregates containing chalk or limestone will be included with the lime content to produce figures that indicate a high percentage of binder. The figures are thrown further into confusion if old, crushed mortar has been included in the mix. The analysis will not identify whether the lime was prepared from chalk or limestone or whether it was site fired in a clamp or factory made. Furthermore, it will not determine the original water: binder ratio or the rate at which it dried. A small batch will need to be prepared and allowed to dry before a comparison can be made; several subsequent batches will probably be necessary using different proportions until a satisfactory mix has been obtained. Careful examination of the broken edge will reveal the number of coats which were originally applied and the thickness of each.

To repair a patch of missing rendering, gently tap around the area to decide where adhesion has been lost and mark it on the surface with a piece of chalk. Using an electrically operated disc cutter, enlarge the patch to remove the loose areas ensuring that the depth gauge is set only to the thickness of the render and at a slight undercut angle all around, except at the bottom edge. Remove the render and form a key on the wall surface by hacking small indentations at about 100 mm centres on a grid basis using a small pick with a handle held close to the face of the wall, to leave a series of indented ledges onto which the render can key itself. Brush down the surface to remove any loose debris and spray it finely with water, using an added fungicide if algae is present. If the clay content of the wall is high, a very fine spray should be used to control excessive movement.

The first coat of render may then be applied by throwing on and trowelling to the same thickness as the original coat ensuring that it is worked well into the undercut around the edge. An indentation is made on the surface to act as a key for the second coat before allowing it to dry. Ensure that it dries slowly by covering with damp hessian sacking and/or occasionally spraying finely with water. Once completely dry, the second coat can be applied in a similar manner.

Trowelling should be kept to the absolute minimum to avoid drawing the fine lime particles to the surface and it should be finished with a wooden float. A third coat may need to be applied to finish flush with the existing render. No attempt should be made to dub out any slight recesses, simply follow the natural contours of the wall to match the existing.

If the wall is to be redecorated externally, the need to match the existing render may not be so important as long as a soft mix is used and continuity is provided by a similar texture. Trial batches based on lime putty, brick dust, crushed chalk and various sands should be mixed and allowed to dry before deciding which is most appropriate. Once the choice has been made, it is wise to incorporate a little animal hair in the render.

RELIEVING STRESSES CAUSED BY UNSUITABLE RENDERS

If a hard, cement-based render has been applied but not yet completely failed, to remove it would cause unnecessary damage to the wall. This

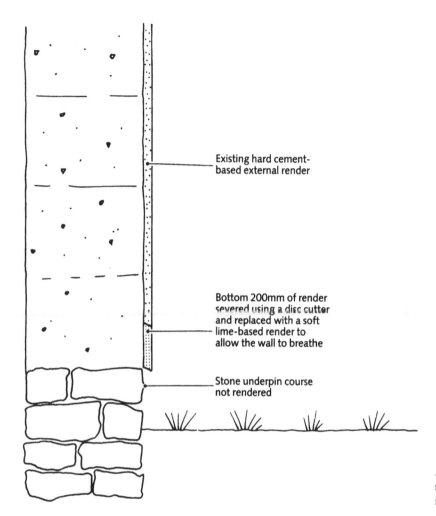

Existing hard cement-based external render

Bottom 200mm of render severed using a disc cutter and replaced with a soft lime-based render to allow the wall to breathe

Stone underpin course not rendered

Figure 62 Relieving the stress of an unsuitable render.

should be avoided. The stresses caused by moisture movement within the wall can be relieved by providing a weak area through which it can breathe. The most likely area would be just above the underpin course where a horizontal strip (minimum 200 mm high) can be removed all around the building. This should be executed with a disc cutter set at an angle to undercut the existing render. The exposed area can than be re-rendered as a patch repair ensuring that a drip is provided at the bottom edge to shed the rainwater clear of the underpin course (Figure 62).

EXTERNAL LIME-BASED RENDERS

Earth structures should not be allowed to dry out completely. A small amount of moisture ensures that the soil particles will stick together and nothing must be done which will inhibit its movement and dissipation. Lime-based renders offer protection to the face of the wall without affecting the porosity of the structure and are recommended. Such renders comprise a bulk filler and a binder in the form of sand and lime respectively but various additives are sometimes employed. The sand needs to be very sharp, angular and well graded to ensure that it bonds together properly. It should be washed and obtained from a river or pit, as sea sand is unsuitable due to the high salt content and rounded shape of the particles.

The lime may be obtained in one of several forms. Traditionally, chalk or limestone was burned in a kiln to produce lumps of quicklime which were added to a pit of cold water on site and slaked. The chemical reaction gave off an enormous amount of heat but once cooled and settled, a layer of lime putty was left on the bottom which was sieved to remove lumps. The resultant soft, white and greasy lime putty was then covered with water to exclude air and kept for at least six weeks to mature.

Quicklime is still available from specialist suppliers and can be used on site today provided that care is taken and protective clothing is worn. However, it is safer to allow the work to be carried out at the manufacturer's premises using hot lime, obtained fresh from the kiln, and to purchase the matured lime putty in sealed plastic bags. Alternatively, it may be obtained in dried form, ground to a fine powder and sold as high calcium white hydrated lime in paper bags through a builders merchant. However, the putty it produces is of inferior quality. One disadvantage of slaking on site is the risk of allowing small particles of unslaked lime to pass through the sieve and be incorporated in the render. These will later 'pop' when in contact with moisture and cause blisters, a process which may last for several years. Such action will react with oil-based paints causing loss of adhesion. To convert hydrated lime powder to lime putty, simply soak it in cold water for a day and ensure that air is excluded. Lime putty prepared in this way is free from 'popping'.

The external face of the wall should be prepared for the render by removing all nails or other ferrous objects, hacking a key as previously described and brushing off loose debris. The surface must then be sprayed finely with water to prevent initial suction.

The lime putty and sand should be mixed together on a flat board, in proportions of approximately 1:3, preferably without the addition of any further water. The sand must be thoroughly clean and sharp and without too many fine particles as these are provided by the lime. Animal hair can be incorporated for the first coat to control shrinkage but the entire mixture needs to be thoroughly beaten to ensure maximum plasticity and ease of application. Supplies of hair from cows, donkeys, goats and horses are available from specialist suppliers. It should be clean, strong but not springy and at least 40 mm long, and needs to be sprinkled evenly over the mix and carefully turned in with a fork to avoid conglomeration into hair balls which result in points of weakness. Exact quantities are not too important but about 5 kg of hair to each cubic metre of render may be used as a general guide.

Ideally, the render should be thrown on, lightly trowelled and finished with a wooden float to a maximum thickness of about 10 mm. Buildings of rammed or puddled earth or adobe should be finished with a flat surface but buildings constructed by the traditional method should follow the natural contours of the wall without dubbing out or finishing with a flat board. It will dry in contact with air but hot sunny weather should be avoided to ensure a controlled process. This can be achieved by spraying finely with cold water for several days and/or protecting with damp hessian sacking, ideally with an open weave to allow air to circulate. Having lightly keyed the surface and ensured that it is dry, a further coat may be applied. Damping down is necessary once again but as the amount of shrinkage is less in the second coat, it is not necessary to add hair.

If the wall is not to be decorated, weathering resistance can be improved if casein, tallow or linseed oil is incorporated in the render. Traditionally, it was more usual to include these additives in the decoration, but undecorated walls were occasionally protected in this way in some parts of the country, as earlier writers have noted. If the lime has been slaked on site, it is wise to add tallow or casein to the slaking process but if it has been prepared from hydrated lime, it is wise to add linseed oil to act as a binder, turning it into an emulsion. The second coat should be of the same thickness, applied by the same method and the drying process controlled in a similar way.

When completely dry, the render should be chemically neutral. This can be tested by wetting a small area with distilled water and pressing on a small piece of pink litmus paper. The render remains alkaline if the paper turns blue.

MAINTAINING THE VERNACULAR

Introduction

The application of an external lime-based render will last for a century or more if applied properly but its longevity should not be the main criterion when considering a form of protection. More research is needed to analyse existing renderings and to classify them both on a regional scale and on a local basis. The visual impact of retaining established local practices should not be under estimated and is recommended as the first option in all cases. The ease of collecting building materials from a merchant is imbued in building trade workers today and the requirement to locate, excavate and cart several cubic metres of clay to mix a mud render is distinctly off-putting and expensive. On being informed of the additional expense, lack of interest in executing the work and lack of longevity, the owner would probably opt for a lime-based render. However, if the vernacular is to be preserved, an effort should be made to seek the cooperation of all parties to reach a sympathetic answer.

Chalk slurries

In the chalk downlands of Wessex, one of the most commonly found types of protection is chalk slurry. Lumps of chalk were broken down, soaked in water and mixed to provide a combined shelter coat and decorative finish. It was applied directly to the face of the chalk wall, presumably following a light spraying of water, but the surface soon broke down so further coats were applied every few years until the thickness built up to 15 mm or more in well protected areas, such as high under the thatched eaves. Where still found, each individual wafer thin coat can be prised apart using a knife. Chalk was dug from the garden, water was available from the stream or well and it became an accepted part of the work cycle for the cottager, farmer and landowner to maintain his property.

Amongst the additives incorporated, limewater was probably the most common which formed a loosely bound distemper. This was obtained by adding excess water to quicklime during the slaking process, allowing it to settle, siphoning off the clear liquid and using it in place of stream or well water. The water contains trace elements of lime which dry and harden in contact with air, thus solidifying the slurry and increasing its durability. There is no historical evidence for the use of pigments and all buildings and boundary walls remain pure white.

Clay renders

In parts of East Anglia, Ireland and Scotland mud renders are traditional. Sand is used as the bulk filler with clay as the binder, mixed stiffly in proportions of approximately 1:3. The inclusion of fibre, usually in both coats, is considered to be essential and hair is more suitable than straw although the latter will suffice if split and chopped. Due to the reduced weathering qualities of the render, it is advisable to add a little linseed oil to improve both durability and binding qualities. Preparation of the wall face is similar to that described for a lime render and two coats, each about 10 mm thick need to be thrown on from a trowel and finished with a wooden float. Damping down with a very fine spray is necessary before each coat to combat suction and it is essential that the render does not dry too quickly. Temporary protection is therefore better than spraying with water to reduce cracking.

Tar

The clay lump buildings of East Anglia stand alone in their use of tar as protection for farm buildings and this tradition is worthy of retention. Tar is a natural bitumen distilled from wood, peat or coal, the last mentioned being the most common, where it was a by-product of the manufacture of gas. Traditionally it was either blended with a sharp sand and applied hot with a brush, or simply applied directly to the clay lump or to a mud render and finished with a sand blinding. Several coats were necessary and it was a simple task to renew it.

Tar is not a waterproof material. It is semi-porous in nature and as the oil evaporates, the tar tends to shrink forming a series of minute cracks through which the wall can breathe. As gas is no longer extracted from coal, coal tar is no longer easily available, but petroleum-based bitumen distilled from oil has taken its place and is available from builders merchants in the form of black bituminous emulsion paint. It is thinner in consistency than coal tar and will not creep but is often used as a modern substitute although the patina it creates does not match the original. Although it is moderately porous to water vapour, the traditional product is to be preferred and the additional effort to obtain it from a specialist supplier is adequately recompensed in the contribution made towards the conservation of the vernacular.

Cement-based renders

Cement-based renders have replaced many traditional renders when a complete renewal has been necessary. These renders are too strong for such a weak substrate and can cause loss of adhesion, rising damp and condensation. No render should be stronger than the surface to which it is applied and although buildings constructed of rammed chalk or stabilized clay are much more dense than their traditionally built counterparts and are able to take a stronger render, those based on lime putty are preferred, particularly if slaked lime is used.

Buildings dating back to the last century exist with their original cement render largely intact, but on analysis the cement content is usually found to be very low. When repairing areas where the adhesion has been lost, a weak mix based on lime putty, crushed chalk, brick dust and sharp sand is advisable. The materials should be mixed and applied with the minimum of trowelling to avoid drawing the lime to the surface and sealing the pores. No case can be made for using metal chicken wire, expanded lathing or ribbed reinforcement. Their use is not necessary and they are not recommended.

Traditional internal plasters

Only buildings intended for human habitation were plastered internally. The basic principle however is different to that of the external render, in that whereas the render protected the wall from the weather, the internal plaster offers an aesthetically acceptable surface solely for the comfort of the occupants. Its only practical use is to prevent dusting and regular removal of debris; the additional insulation provided and the ability to hide electrical cables are advantages which have been inherited by the recent generations of owners.

There are fewer regional variations in internal wall finishings, and a two coat system is usual with a three coat system reserved for the better quality buildings of rammed earth construction.

In a two coat system, the first, or render coat is thicker than its external counterpart, often up to 20 mm, and based on lime putty and sand (1:3) with hair worked in. The final or setting coat is thin, rarely exceeding 3 mm and is either of pure lime putty, or of 'plasterers' putty' composed of lime putty and silver sand in equal proportions. Both coats are applied and finished with a wooden float and do not contain any additives as the need to improve durability against the weather does not arise. To include an additive on both sides of the wall will inhibit the free movement of moisture and can cause damp to rise.

In a three coat system, the render coat is usually about 10 mm thick and incorporates hair. The second, or floating coat, is similar but maybe without the hair, and the setting coat is a thin layer of neat lime putty. The overall thickness is about the same as a two coat system.

In some parts of the country mud plasters have been used, sometimes as a single coat of clay and sand about 20 mm thick and sometimes as a two coat system where a 15 mm render coat is finished with a thin setting coat of lime putty. Variations on this theme include a two coat system with a render coat of clay and lime and a setting coat of sand and lime applied before the render coat has dried.

Chapter Seven

Wall decoration

INTRODUCTION

Having ensured that a porous render has been applied, it is folly to seal the surface either internally or externally with a paint film or fabric which will prevent the natural flow of moisture through the wall. All decorative materials must therefore be porous.

Where a chalk slurry has been applied directly to the face of a chalk wall, it acts as a combined render and decoration although its life is short and regular renewal is necessary. Where a clay slurry or lime render has been applied to a clay wall, its function is purely protective but colour washes have usually been added, both for decorative effect and to protect the render. The most suitable form of decoration is one which is both compatible with its substrate and water permeable allowing damp within the wall to evaporate. Such paints also reduce problems caused by condensation. Lime-washes and soft distempers are the traditional materials and are to be preferred for the sake of the structure. However, potassium silicate and cement-based paints have now been in use for a century and have often been used, although they are not as compatible with earth structures as limewashes and distempers. Modern masonry paints have been developed which are moisture permeable to some extent and these have been used by the less traditional minded owner, although they usually lack character and fail to develop a patina sympathetic to the building. It is important that they are not oil-based which may cause saponification with a damp, lime-based substrate. Saponification occurs when the oil becomes soft and sticky, turning into a soapy mass and can only be rectified by complete removal.

The use of silicone damp proofers is not recommended for use on any earth building. They are manufactured from silicone resins which are polymers containing silicon, and are sold as clear water repellants but they are not completely waterproof.

If an existing render or limewashed surface is being redecorated, any sign of algae, moss or lichen should first be destroyed by using a biocidal wash. The problem is more likely to arise on an existing render as there is evidence to suggest that new limewash will inhibit mould growths, particularly if it contains iron sulphate.

A biocidal wash is prepared from a concentrate which can be diluted in water. Heavily contaminated areas should be scraped to remove thick growths before local treatment with the undiluted concentrate by brush or spray. Areas suspected of contamination can be treated with a 1:3 dilution in water as a general wash and left to dry. Up to seven days is needed to kill fungal or bacterial spores but the wash will remain active for up to two years. Dead growth should be brushed off before the limewash is applied.

A paint is a pigment suspended in a liquid. The pigment provides opacity and colour and the liquid comprises a thinner and a binder. The thinner is usually, turpentine, white spirit or water and the binder may be bitumen or cellulose derivatives, resins, varnishes or various oils. Paints dry by different methods depending on their composition but usually by oxidation or by evaporation of the solvent. Dryers are sometimes added which work by chemical reaction. In modern masonry paints additives may include extenders (bulk fillers), water-proofers, accelerators (to hasten drying) and aggregates to give texture.

A water paint is one which is capable of being thinned by water, although its binders may be insoluble. By this definition, modern acrylic resin-based paints are water paints. Unpigmented limewashes and soft distempers are generally referred to as whitewash.

TRADITIONAL COATINGS

Limewashes and additives

Traditionally, the most commonly found decorative medium both internally and externally is limewash. However, its recipes vary considerably producing washes with a wide range of characteristics, depending on the additional ingredients included.

In its simplest form, limewash is lime putty to which more water has been added to turn it into a milky consistency for application by brush. Alternatively, bagged hydrated lime may be soaked in water for a day, mixed with emulsifying additives and thinned ready for use. When used externally it requires regular renewal but the addition of a binder such as casein, tallow or linseed oil will improve its durability and should always be considered essential, particularly if the limewash has been prepared from hydrated lime.

If limewash is made from hydrated lime, an inferior product will be produced which will rub off easily and be short lived. This is because the particles have been dried and ground finely, offering a greater surface area which carbonates on exposure to the atmosphere, resulting in reduced binding qualities. Limewash made from lime putty has not been dried and ground and will not carbonate until it is applied. It is therefore recommended that all limewashes are made from matured lime putty.

Casein is an animal derivative which was once available from dairy farms. When hydrochloric acid was added to milk it reacted chemically allowing the coagulated protein to be skimmed off the watery whey and dried. Today, it is sold in packs of 500 grams and can be added to a limewash to act as a water repellent and to prevent dusting.

Tallow is the refined fat of a cow, pig or sheep and serves a similar function to glue size but is more commonly found. As a binder and weatherproofer, the life of a limewash is extended considerably and tallow is obtainable in solid blocks weighing 1 kg. It is used like casein.

Linseed oil is a traditional vegetable-based additive which is used as a binder and water repellent. Its water shedding properties are considerable and lime renders or limewashes need only a small quantity to ensure a much longer life before a further coat is required. The oil is obtained from the seed of the flax plant (*linum*) which has experienced a rebirth in the countryside in recent years where fields of pale blue flowers dominate the chalk slopes in early summer. It is obtainable from specialist suppliers in its raw, clarified form where it can be whisked or beaten into a limewash or lime putty.

If limewash is being prepared from quicklime, the casein or tallow should be incorporated in the slaking process to ensure that it melts and is involved in the chemical conversion. A 10 per cent solution (i.e. 1 kg to 10 kgs of quicklime) is a general guide but old recipes vary from half to double this quantity. It should be borne in mind that a greater quantity of additive will not only increase the protection against the weather but inhibit moisture movement within the wall. If the limewash is prepared from hydrated lime, the casein or tallow needs to be dissolved in hot water and beaten in thoroughly after the lime putty has been thinned (although casein will need a little lime to help it to dissolve as it is insoluble in water but soluble in alkalis). This is not as satisfactory as chemical conversion and linseed oil may therefore be preferred, a 1 per cent solution being all that is required (i.e. 50 ml to 5 litres of wash) well whisked into the diluted lime putty.

In the past, various additives have been used including wood ash, cement, flour, buttermilk, sulphate of zinc, alum, calcium stearate, keratin (to retard drying), copper sulphate (as a mould inhibitor), rock salt, trisodium phosphate (to assist the dissolving of casein), milk, Fairy soap flakes, beeswax, slag from iron smelting and doubtless many others not recorded. Sometimes several additives were used in combination to produce a wash with special characteristics to suit a particular situation.

Most additives improve durability by reducing porosity, and salt has often been incorporated to counteract this tendency. For a wall as porous as one built of earth, the addition of salt is not recommended as it will be inherent in the structure and to add to it may cause injury.

Limewash is ideal for internal use as it is unaffected by rising damp which continually breaks down most other paints. However, it is not wise to incorporate weather resistant additives such as casein, linseed oil or tallow as this will reduce absorption and hinder the evaporation of moisture within the wall. If a little alum is added (25 grams to 5 litres of wash) it acts as an excellent binder and prevents it from rubbing off.

Limewash is not ideally suited for application over impervious surfaces such as flint or granite underpin courses, cement-based renders or modern non-permeable paints, but in East Anglia, buildings are known to have been protected with coal tar followed by limewash applied while the tar was hot. The underpin course is better left unprotected, both to ensure that the amount of moisture rising into the earth wall is kept to a minimum and from an aesthetic point of view. However, if it is protected with a soft render, the application of limewash will do no harm.

An existing limewash should always be redecorated with a new limewash. Manufacturers of paints recommend sealers for porous surfaces prior to the application of their masonry paints but they should be chosen with care as their moisture permeability qualities tend to vary. If for some reason, it is considered necessary to apply a modern masonry paint over an existing limewash, the entire wash should be removed to expose the render before applying the new decoration, rather than applying a sealer.

To remove a limewash is not an easy task and is not recommended; intensive action can cause the loss of the render. Wetting the limewash and scrubbing is the only safe method of removal but considerable labour is necessary. Chemical paint strippers are not recommended. To redecorate a limewash with a new limewash is always preferred.

A new render needs to be well dampened and the first coat applied very thinly by brush, working it in thoroughly to ensure that the surface is fully covered. The surface must be saturated with the wash which will appear translucent, as the pigment will only appear as it dries. If applied directly to the earth face, as was often the tradition in Ireland, the wash will form a slurry with the earth which will crack as it dries out. Subsequent coats usually obliterate the problem. Temporary protection from rain or sun is necessary until the first coat has dried, before damping down and applying the second coat. A minimum of three coats internally and four coats externally should be used as a guide, with each coat applied as thinly as possible. Too many coats will obliterate detail. Each coat should be applied to a small area at a time, pre-damping as the work is executed.

The life of a limewash is dependent on the substrate, the formula, the method of application and the degree of exposure. A life expectancy of twenty years can be achieved when all are in harmony.

Distempers and water paints

Distempers and water paints, like limewashes were always mixed on site, in just sufficient quantity for each task, any leftovers were discarded. Chalk was crushed, washed, dried and ground to form whiting which was used as a paint extender, as the bulk filler for glazing putty and as the basic white pigment for distemper and other paints. Whereas limewash was the common external decorative finish, distemper and water paints were the common internal decorative finish and were made by dissolving a casein oil or size binder in the whiting, adding a pigment and diluting to a thin cream consistency. Their life was generally short, although some distempers survive of great antiquity. In better quality work, white lead was used in place of whiting.

Size is a weak, transparent, liquid glue which was used as a sealer of porous surfaces to prevent paints from soaking in. There are two types. Varnish size is made from resins and oils which are thinned and used on timber surfaces, and glue size is made from animal bones, horns and hoofs (keratin) which are thinned with water and used on plastered surfaces. Glue size is available from specialist suppliers and may be used as a primer for distemper but not all authorities recommend its use as it is prone to flaking on damp plaster.

Two types of distemper were in common use. Soft distemper in powder or paste form comprised whiting, pigment, and a binder of glue size. Hard distemper contained soap which made it a much better product. If the distemper had a binder of casein or linseed oil, it was known as an oil bound water paint which was vapour permeable but likely to be affected by alkalis. Sometimes, a combination of binders was added and driers were incorporated if circumstances dictated their use. They were also used externally.

A whitewash can be made by putting 3 kgs of whiting in a large bin together with a nugget of alum, covering this with cold water and leaving it to soak for 30 minutes. The excess water is poured off and a thoroughly soaked powdered pigment added and beaten well in to form a thin batter.

A soft distemper can be made by putting 3 kgs of whiting in a large bin, covering it with hot water and leaving it to cool. The excess water should then be poured off and a well soaked pigment, together with 1 kg of glue size, whisked in and diluted.

In warm weather, a little glycerine or flour paste can be added to soft distemper to delay drying.

If the substrate is extremely porous, it can be solidified by dissolving 500 grams of glue size and 50 grams of alum in 5 litres of hot water and leaving to cool before applying with a brush. Although distempers are compatible with lime-based renders, lack of experience will make it difficult to apply them to some modern 'renovating' plasters.

Traditional distempers are easily disfigured by damp walls and mildew growth will soon appear. Affected areas cannot be touched up to provide a satisfactory surface and when redecoration is required, a much better finish is obtained if the previous distemper is completely removed.

Earlier this century premixed distempers became available which were capable of being diluted further but their use has been largely overtaken by emulsion paints. Distempers are available either in 5 or 10 litre pails as a non-washable soft distemper for internal use, or with an oil base in 5 litre cans as a scrubbable water paint for internal or external use where a high vapour permeability is required. The latter is particularly difficult to remove.

Cement-based paints

Cement-based paints provide a cheap, durable, water repellent decorative coating which is compatible with lime-based renders. They are suitable for porous surfaces but produce a hard, matt finish. Originally they were mixed on site from Portland cement and water to which a binder, pigment and extenders were added.

To make a cement paint, place 10 kgs of Ordinary Portland Cement in a large tub and mix with a little water to form a slurry. Mix together 1 kg of hydrated lime and 1 kg of plaster of Paris (hemi-hydrated gypsum plaster) and blend it with the slurry. Melt 1 kg of glue size together with the pigment in hot water and add it to the mixture with a little alum. Mix well together and dilute. Two coats are normally necessary but the wall should first be dampened. Ideally, it should dry slowly and application in hot sun is not recommended. The paint will last many years until the top coat chalks and dusts away when it can easily be replaced by another coat. Pastel pigments are more suited than strong colours for fear of a white bloom developing.

Cement paints are heavy and the weight increases each time another coat is applied. On porous surfaces the paint may pull away once the thickness builds up with successive coats over the years and the only remedy is to remove the entire coating and start again. Being less flexible than the substrate, they are prone to cracking thereby allowing moisture to enter and salts to accumulate.

Snowcem is a premixed cement-based paint sold in powder form and requires only the addition of water. It is water repellent, moisture permeable and conforms to the British Standards Specification 4764 (1986) for powder cement paints for use as decorative and protective coatings for internal and external use on porous surfaces. It is made from white Portland cement together with waterproofers, accelerators to hasten drying, and extenders to ease application. Snowcem should not be applied over limewashes or emulsion paints, nor should it be applied when rain is due or in cold weather. If committed to a cement-based system it should be maintained as it tends to be incompatible with most other paints.

Cement-based paints have been used to provide external protection to earth buildings for many years. Their rate of permeability is not high and this is reduced with each successive coating, making them a poor substitute for limewash.

Pigments

Limewashes and distempers are white in colour and cement paints vary within light shades of grey, but pigments may be added if a tint is desired. At one time, coloured clays are thought to have been used to produce pastel shades in the west country and a finely ground brick or stone dust was used in the eighteenth century to provide permanent tints.

Pigments are either natural or chemically prepared. Natural pigments are obtained either from organic sources (i.e. from plant or animal origins) or inorganic sources (i.e. from mineral origin) the former being variable in lightfastness and the latter generally permanent and dull, apart from a few exceptions.

Natural pigments are based on oxides or hydroxides of iron and are usually mixed with clay or silica. Whiting is the most commonly found pigment for white but French chalk (talc), china clay (kaolin), ground quartz (silica) and barytes are also used. Reds and browns are derived from iron oxides or umbers, blues and greens from verdigris and indigo, yellows from ochres or siennas and blacks from iron oxides, manganese oxides or graphite.

Chemically prepared pigments have been developed in response to a demand for a more reliable colour consistency. Their light fastness may be variable in some colours but most of them have outstanding qualities and uniformity can be guaranteed. Calcium plumbate, titanium dioxide, lithopane, white lead, zinc phosphate and zinc oxides are all used to provide white pigments. Reds and browns are produced from cadmium red, organic red, red lead or synthetic iron oxides, blues and greens from Prussian blue, chrome green or chromium oxide, yellows from zinc chrome, lead chrome, organic yellows or iron oxides, and blacks from carbon, vegetable black or iron oxides. However, Prussian blue is not stable in lime or chalk mediums.

Pigments are sold in powder, paste and liquid form. In powdered form they are sold in 0.5 kg and 1 kg bags or in 3.5 kg plastic containers in a wide range of clear, brilliant colours which can be blended together to give subtle tints. They should always be soaked well in the medium to ensure thorough dispersion before being added to the paint. If added dry some particles may not dissolve, resulting in streaking. In paste form they are sold in 50 ml tubes of concentrate but it is still advisable to soak them in water and to mix them into the medium thoroughly. Organic, mineral-based pigments in liquid form are sold in cans containing one American pint (approximately 500 ml) and are available with a base of either water or linseed oil.

The addition of a pigment needs to be considered carefully as it is not possible to judge its final colour until a trial mix has been tested and allowed to dry, as most colours tend to look very much darker when wet. Several trials are usually necessary to achieve the desired tint and a careful note should be made of the quantity of pigment added to each trial mix. Once satisfaction has been obtained, sufficient limewash for the entire wall should be tinted to prevent variations which might arise if several batches are tinted individually.

The amount of pigment to be added depends upon the intensity of the shade required. In powdered form this varies from 400 grams to 5 litres of limewash for the light shades to 1 kg for darker shades. It is not wise to add too much pigment to a limewash as its binding qualities will be reduced but the quantity of binder can be increased slightly to counteract the tendency.

British Standards Specification No. 1014 (1977) has been produced for pigments but these refer only to those used in conjunction with

Portland cement. They are readily available from builders merchants for use in mortars, renders and concretes and contain synthetic oxides combined with accelerators and plasticizers. These pigments are sold in powdered form but when added to a cement-based product its permeability is reduced to such an extent due to their microscopic, even particle size, that it becomes almost waterproof. The use of such pigments is not therefore recommended.

Pigments suitable for use with limewashes and cement paints can be obtained from specialist suppliers to the trade and from art and theatrical shops.

MODERN COATINGS

Introduction

The trend from paints mixed on site containing basic ingredients to factory manufactured products supplied ready for use has enabled consistency to be achieved and quality to be maintained. It has also encouraged the development of new products which have largely taken over from traditional forms of decoration and protection. Emulsion paints have replaced distempers and masonry paints have replaced limewashes for modern buildings.

The advantage of modern paints lies in the control of their manufacture and the technical assistance offered by the manufacturer in the event of any problems which might develop. The manufacture of modern paints is a complex process which uses the wide range of materials available to the industry to produce a high quality product which is scientifically tested before being marketed. Never before has the customer been offered such a wide range of products.

Most paints are moisture permeable to some extent but it is a two-way process. Those which repel rain effectively are low in permeability and those which offer only a low protection against rain are high in permeability. The wide range of masonry paints now available enables the customer to select the product which gives him the degree of permeability/protection he desires.

In recent years, considerable effort has been made by the marketing companies to stress the permeability of paints, describing them as 'microporous'. It should be borne in mind, however, that unless the degree of permeability is stated and compared with others, the word has little meaning. British Standards Specification 3177 (1959) sets out the criteria for the measurement of 'Permeability to water vapour of flexible sheet materials'. It was prepared for use by the packaging industry but has been adopted by most of the paint manufacturers and serves as a standard by which comparisons can be made. Moisture permeability is measured in the number of grams of water which will pass through one square metre of paint film in one day. This is dependent upon the temperature, the relative humidity and the thickness of the dry paint film, which is usually stated in microns. Paint manufacturers are helpful in providing results of moisture permeability tests and they are often published as part of the technical specification. Claims of a product being 'high' or 'low' should be regarded with suspicion unless figures can be produced and compared.

Compatibility with the substrate to ensure proper adhesion is just as important as permeability. Some types require all traces of previous paints to be removed, whereas others are compatible provided the existing surface is primed. The choice of primer is just as relevant as the choice of paint, and moisture permeability test results should be obtained from the manufacturer before application of what is most probably a permanent coating. Their use should therefore be avoided if at all possible.

Loss of adhesion is usually caused when a paint film is not sufficiently permeable to allow a damp substrate to shed its moisture. Rising damp carries salt nutrients with it which crystalize on the surface when the moisture dries out, pushing off the paint film. Even where salts are not present, chemical reaction between paint and substrate may fade the colour of some pigments if the substrate is constantly too damp.

Modern paints have been used on earth buildings for many years, often with a mixed degree of success. The qualities of limewash and distemper cannot be matched in terms of porosity and texture, and as these are crucial to the wellbeing of the structure and to the preservation of the vernacular tradition, they are to be preferred.

'Pozament' limewash

Limewash is now available premixed in a hydrated form and sold in sealed bags weighing 10 kg. Known as 'Pozament Polymeric Limewash', it is a mixture of hydrated lime and 10 per cent pozzalanic pulverized fuel ash (cenospheres) together with a polymeric binder as a thickening agent. The powder is added to water and mixed until it reaches the consistency of 'pouring cream' when it is ready for internal use. For external use, a little linseed oil is advisable (50 ml to 10 litres of wash), well whisked in, but tallow and casein are not recommended due to the inclusion of the polymeric binder. Pozament is applied in exactly the same way as a limewash prepared from lime putty but if applied directly to the face of an earth wall, it is wise to apply a coat of size or petrifying liquid first. When applied over distempers, whitewash or thick limewash, problems with adhesion may occur and the removal of the old coatings is therefore recommended. Three coats are usually adequate.

Internal emulsion paints

An emulsion paint is one in which particles of a liquid or solid medium and a pigment are suspended in water. The most commonly found mediums are bitumens, oils and synthetic resins. When applied by roller, brush or spray, the particles are broken and dispersed and run together to cover the surface completely. Being a water paint, emulsions are non-flammable, porous, can tolerate a certain amount of damp, and dry by allowing the water to evaporate leaving behind the pigment and the medium which now acts as a binder. Such paints are considered 'non-convertible' as opposed to 'convertible' paints which dry either by oxidation (i.e. by absorbing oxygen from the atmosphere) chemical cure (i.e. two part packs which dry once mixed together) or by stove enamelling (i.e. by the application of heat).

Modern emulsions used in the building industry are based mainly on solid synthetic vinyl resin particles suspended in a water solvent and they produce varying degrees of permeability. Polyvinyl acetate, and acrylic resins are the most common types but copolymers of PVA and polystyrene resins are also used.

PVA emulsions are odourless and porous to water vapour depending upon the type of finish. Gloss and eggshell finishes are reasonably water permeable whereas semi-gloss finishes tend to be less water permeable. They dry quickly, to provide a washable surface and are alkali resistant. Acrylic emulsions are more highly porous and dry quickly to a flexible, washable finish but are more expensive than vinyl. They have a greater adhesion and tend to produce a better finish on absorbent surfaces but give off a characteristic smell during application. Being non-oil-based, they are alkali resisting. Acrylic-based emulsion paints are suited to the internal decoration of earth buildings and some are suitable for external use, but they do not match the qualities of traditional coatings.

The qualities of vinyl copolymers and polystyrene resins are considered in the section for masonry paints.

External masonry paints

Most masonry paints manufactured in this country are non-convertible emulsions based on synthetic acrylic resins with added grit to improve durability or to create a textured finish. However, vinyl copolymers and polystyrene non-convertible resins are also used as well as convertible mediums such as drying oils and alkyd resins which may be thinned with white spirit and dry by oxidation. Masonry paints containing a medium of drying oils are not recommended for use on lime-based renders to earth buildings due to the risk of saponification. Epoxy resin-based paints in a water solvent are also available in two part packs which cure chemically but these are almost waterproof and are not suitable for buildings which need to breathe. Blends of two or more synthetic resins produce paints with distinct characteristics which are currently the subject of research.

The properties of masonry paints are determined by the liquid medium and the solid matter suspended in it. They should not be thinned as the distribution of the solid matter is upset and this affects the performance of the paint. The nature of the grit varies from one manufacturer to another, mica, quartz, granite and sand being the most commonly found.

All masonry paints are intended for external use and provide a durable coating offering a high degree of protection in a wide range of finishes from high gloss to dull matt. A wide range of textures is also available from a fine, smooth surface to rough stipples, but few of them are able to match the character of traditional coatings.

The non-convertible acrylic-based emulsions have already been considered but when used as a masonry paint, their porosity is dependent upon the dry film thickness which varies considerably from one manufacturer to another due to the inclusion of different types of grit. Being water paints they must not be applied in wet weather for fear of being washed off, or applied in cold weather below 5°C as they are easily damaged by frost. They do not penetrate the substrate very deeply and primers are often necessary to ensure good adhesion, particularly on porous surfaces. Being of a thermoplastic nature, they tend to pick up dirt, particularly if they have a textured surface. The relative humidity tends to dictate the drying time but in normal conditions they dry fairly quickly. Some manufacturers include a mould inhibitor while others recommend a separate application of biocide wash. As with other water paints, they can be applied to a damp surface but if affected by rising damp, some acrylic-based masonry paints may react chemically with the salt nutrients.

'Pliolite' is a trade mark registered by the Goodyear Tyre and Rubber Company of Akron, Ohio, USA. It is a convertible styrene-butadiene resin binder and offers a different range of characteristics to those paints based on acrylic emulsions due to its particles being much smaller. Despite the fact that the paint film is normally thinner, Pliolite offers a much lower rate of moisture permeability, although one manufacturer makes a paint which is so thick that it can be applied by a trowel. Pliolite-based paints are not water paints and need to be thinned with white spirit. They are flammable, with a flash point above 32°C and tend to produce a rather unpleasant odour during application. Manufacturers claim a 10 year life before renewal due to their good alkali resistance and high qualities of adhesion by penetrating the substrate, usually without a separate primer. As the solvent is a spirit they can be applied in cold, unstable weather down to minus 10°C, but with difficulty. Most incorporate a mould inhibitor. Their use is not recommended on earth buildings.

Alkyd resins are made from fatty acids derived from vegetable oils such as soya, linseed, sunflower, etc. They form the largest group of convertible, synthetic resins used in the paint industry and although sometimes used as a base for emulsion paints, their use in masonry paints is limited. Their use offers ease of application but they are easily affected by humidity while drying. Alkyd-based paints are unsuitable for damp surfaces and are not therefore recommended for use on earth buildings.

Vinyl copolymer resins are non-convertible resins which are used as a binding medium in a few masonry paints. They are water thinnable and highly porous to moisture vapour despite the thick paint film they produce. Although capable of being applied to damp surfaces, they should not remain in contact with moisture and their use is not therefore recommended.

Several manufacturers offer colourless water repellents for enhancing the weather resistance of their masonry paints. Such products have a very low rate of moisture permeability but they are not waterproof. They should not be used on any building which needs to breathe.

The range of masonry paints available offers the customer a wide choice of qualities. For all normal circumstances, those based on non-convertible acrylic emulsions are the only ones likely to be suitable for earth buildings but if any doubt exists, the technical service offered by the manufacturers should be consulted.

Silicate paints

Paints based on sodium or potassium silicate have been manufactured for over 100 years. Although no longer produced in the United Kingdom, they are still made in Germany and are available from the importing distributor but tend to be rather expensive.

Silicate paints comprise mineral fillers with an algicide and inorganic pigments in a binder of potassium silicate (waterglass). They are not water paints and differ from other products in that they penetrate the substrate and form a chemical bond with it (silification) rather than applying a film over the surface. This ensures a very high rate of moisture vapour permeability making it suitable for buildings whose porosity needs to be maintained to ensure their survival.

Several types of paint are imported, but the emulsion for internal use and the masonry paint for external use are those most likely to be encountered. Both are alkali resisting, odourless and present a smooth, matt finish but neither should be applied in temperatures below 5°C. It is essential that the unique silification qualities are not impaired so all previous paints need to be removed and highly absorbent surfaces primed with non-pigmented liquid potassium silicate to ensure long lasting protection. Tests made by the British Board of Agrément indicate a minimum life of 15 years, which is longer than for most other masonry paints and compares favourably with a well applied limewash.

The emulsion paint is sold in paste form for dilution on site with liquid potassium silicate whereas the masonry paint is available either in powder form for site mixing with powdered pigments and a water-borne potassium silicate solution, or in liquid form which requires further dilution on site with a non-pigmented liquid potassium silicate.

Recent research in the use of potassium silicate for the consolidation of wall paintings on lime plasters has suggested that slight carbonation can take place giving rise to superficial efflorescence and casts doubt on the use of the chemical for this type of work.

However, if a silicate paint system has been introduced, it is important to maintain it to ensure that its outstanding porosity and silification properties are not impaired.

PERMEABILITY TABLES

A guide to the relevant moisture permeability of traditional and modern paints appears on the following page in tabulated form.

WALLPAPERS

If a patterned decoration is desired for an internal wall surface, it should be chosen with care. It is important to ensure that the porosity is maintained and that algae and fungal growths are not encouraged.

Paper is made from a mixture of cellulose and water which is rolled into a thin sheet and allowed to dry, although additives are used to produce high quality papers. Linen and esparto grass are sources of cellulose to the industry, although wallpapers are usually made from 80 per cent mechanical wood pulp and 20 per cent unbleached sulphite pulp and printed on one side.

Type of paint	Binding medium	Drying medium	Indication of comparative moisture permeability
Limewash	Tallow, casein or linseed oil	Evaporation	Very high
Cement paints	Cement	Evaporation	Medium
Soft distempers	Glue size	Evaporation	Very high
Water paints	Linseed oil or casein	Evaporation	Variable
Emulsion paints	Bitumen or tar	Evaporation	Medium
Emulsion paints	Polyvinyl acetate	Evaporation	Variable depending upon finish
Emulsion paints	Acrylic	Evaporation	High
Masonry paints	Acrylic	Evaporation	Variable but generally high
Masonry paints	Vinyl copolymer	Evaporation	High
Masonry paints	Alkyd	Oxidation	High
Masonry paints	'Pliolite'	Oxidation	Low
Masonry paints	Epoxy resin	Chemical cure	Very low
Silicate paints	Potassium silicate	Oxidation	Very high

Wallpapers are highly porous and will not inhibit the movement of moisture within an earth wall provided that they are not applied with epoxy-based adhesives. A vegetable starch-based paste in a water solvent is recommended in preference to a PVA paste or glue size and those which contain a fungicide will ensure freedom from mould growths. British Standards Specification 3046 (1981) provides for five types of adhesives for hanging wallpapers. Types 1, 2 and 3 cater for papers of different weights and Types 4 and 5 cater for different strengths of sheet vinyls and include mould inhibitors. Moisture permeable starch-based adhesives are available with mould inhibitors which comply with Types 1, 2, 3 and 4.

Embossed papers are highly porous, cellulose-based, bleached coverings on which a design has been impressed as opposed to printed. They are ideally suited for use on earth buildings and may be finished with a pigmented acrylic emulsion paint if desired.

Washable coverings are available in two types. Wallpapers coated with a thin film of vinyl have their porosity considerably reduced but are still moisture permeable to a similar extent as vinyl-based emulsion paints. Sheet vinyl coverings, however, comprise much thicker, solid synthetic resins which are almost waterproof. They are not based on cellulose paper and their use is not therefore recommended.

Chapter Eight

Conclusion: the future of earth building

The use of earth fell into a slow decline in the last part of the nineteenth century, finally ceasing early in the twentieth century in most areas. Since then, only occasional experimental work has been carried out. However, there have been several calls to return to earth building, usually at a time of national crisis such as during oil supply shortages or after the two world wars when the building stock was in need of replacement as a matter of urgency.

Perhaps the best known voice advocating such a return in this country belonged to the architect Clough Williams-Ellis following the publication of his book *Cottage Building in Cob, Pisé, Chalk and Clay*, in 1919. The introduction was written by St Loe Strachey, his father-in-law and editor of the *Spectator* magazine, in which he stated that:

> The country is faced by a dilemma probably greater and more poignant than any with which it has hitherto had to deal. It needs, and needs at once, one million new houses, and it has not only utterly inadequate stores of material with which to build them, but has not even the plant by which that material can be rapidly created.

Williams-Ellis gave the lead by designing and constructing several buildings upon which he reported when the book was revised, enlarged and republished in 1947. In the Preface, he stated that:

> So similar are building conditions in this spring of 1947 to those obtaining after the first world war twenty seven years ago, that the introduction then written by my late father-in-law has re-tained exactly as he wrote it as being still entirely and disturbingly appropriate.

The call met with an indifferent response in countries throughout the world. In the British Isles, the call was largely ignored but he recorded that he was 'particularly surprised' at the interest shown by the Scandinavian countries and by Canada. The prominent architects of the day rose to the challenge, always anxious to experiment with a 'new' material. Le Corbusier designed earth buildings in France, Schindler and Frank Lloyd Wright in the United States and Lutyens in the United Kingdom, but, apart from one of Lloyd Wright's houses, they failed to materialize.

In the United Kingdom, it is claimed that the introduction of the Building Regulations prohibits the constuction of earth buildings. In the past, this was true in those towns which had adopted the Model Bye-laws, as Dr Poore found out in 1901 when he planned to build his chalk cottage in the Borough of Andover (Hampshire). The Bye-laws stated that the walls of a house must be of hard, incombustible material bonded together by good mortar or cement. As chalk is not hard, contains straw and is not bonded to anything, he was not allowed to build the cottage within the Town Board limits, much to his annoyance. In defiance, he built it just outside the boundary at Gallaghers Mead where it stood until its demolition in 1985, to be

Figure 63 The delightful chalk cottage built by Dr Poore in 1901 just outside the town boundary of Andover (Hampshire) in protest against the Building Bye-laws. The building set new standards in rural sanitation. It was demolished in 1985.

replaced by modern housing (Figure 63). Other examples of applications to build earth buildings which were rejected are known as late as the 1920s.

The Building Regulations have been rewritten and revised several times since then and are now mandatory throughout the country. A degree of flexibility has been introduced which allows a Local Authority to approve the construction of earth buildings, subject to any reasonable conditions the Authority may require. Approval has been given to several applications subject to tests being made on site to establish strength, moisture content, etc. The Regulations do not apply to all types of structures and those that are exempt may be constructed in earth subject to planning consent. Boundary walls, bus shelters, storage sheds, cottage extensions and houses are all known to be planned, approved or built in recent years and two examples are shown in Figures 64 and 65.

Figure 64 A boundary wall of reconstituted chalk at Andover (Hampshire) rebuilt by the author in 1983 by the traditional method. One course was built each week for ten weeks and the paring was executed as one final operation. One year later, the wall was rendered and decorated using tinted Pozament polymeric limewash.

Figure 65 The bus shelter at Down St Mary (Devon). The structure was built by Mr Alfred Howard in 1978 of clay cob off a high underpin course of worked stone. The walls were protected with a cement-based Tyrolean render and decorated.

Such schemes are small and most effort is being directed towards maintaining the existing stock of earth buildings rather than constructing new ones. In south-western England the Devon Earth Building Working Group was formed in 1991 to provide a forum for discussion, give advice on maintenance, organize seminars and demonstrations, prepare and publish a repair manual, train local craftsmen in repair techniques, encourage building in earth and to forge links with other groups having similar interests. The group comprises representatives from English Heritage, local authority building control officers, the Building Research Establishment and the Devon Rural Skills Trust together with local representatives from the National Trust and the Society for the Protection of Ancient Buildings. It also draws on the experience of craftsmen who have gained experience on local repair projects. The first paper was published by the group early in 1992.

Some countries returned to earth after the Second World War to help them provide accommodation for the returning troops, to cater for the population explosion and to replace their bomb damaged buildings. More than 300,000 were built in Eastern Europe to alleviate the shortage, most of these in Germany. Similar circumstances occurred in Korea after the armistice was signed in 1953, and rebuilding was supervised by Eastern Europeans.

Adobe buildings have been associated with New Mexico for many centuries and in this quiet, south western corner of the United States this method of building has never died out. Quite the contrary, recent developments in the state have been spectacular with modern, light, airy buildings being constructed on such a large scale that a regular monthly magazine *Adobe Today* has been published dealing exclusively with adobe design and construction. Possession of such a house has become a status symbol in recent years.

In 1942, the French architect Le Corbusier published 'Les Murondins' (as quoted in Dethier's 'The Story of Pisé') in which he called for a return to earth and justified both the use of the material and the part it should play in the post-war reconstruction programme. In 1947–8 he designed housing projects for a city near Marseilles, houses which, unfortunately, were never built.

In 1970, 'The Centre for Research and Application of Earth Construction' was founded in France. Known as CRATerre, its sole aim is to modernize the use of earth for building purposes both in Europe and the Third World. It was founded by a group of engineers and architects within the School of Architecture at Grenoble and, working in conjunction with Grenoble Scientific University, it has established itself as a world leader in earth technology. It has advised on projects in 30 countries during the last 10 years on behalf of the United Nations and World Bank.

The fuel crisis of the 1970s brought a further call to return to earth. This was heralded by the exhibition 'Le Génie de la Terre' which opened at the Georges Pompidou Centre in Paris in 1981 under the direction of architect Jean Dethier. Its main aim was to demonstrate the feasibility of earth as a building material of the future in addition to showing the wide range of earth architecture to be found throughout the world. The success of the exhibition was followed by a world tour where it was seen by over three million visitors and the accompanying catalogue was translated into seven languages.

To demonstrate the feasibility of earth architecture, a new town is being built between Lyon and Grenoble. Known as L'Isle d'Abeau, it is an experimental town on a site of 2.2 hectares, originally of 40 houses but now expanded to 63. The town was planned by CRATerre with the needs of the Third World in mind and the accommodation includes buildings up to 5 storeys high. About 10 per cent of the buildings are of timber frame construction with earth walls built by the traditional method and finished with a lime-based render. The remaining 90 per cent is shared equally between buildings of rammed earth and of stabilized adobe blocks.

The construction industry in the United Kingdom consumes about one third of the country's total energy requirements, a figure typical of most industrialized nations. Earth architecture for certain classes of building is claimed to reduce that figure, with savings on the manufacture of machine tools, the firing of clay and the transport of materials. Pollution of the planet is claimed to be reduced and the high insulation qualities of earth ensure the economic use of the world's energy resources. It is thought that a reduction in construction costs would follow in due course aided by the use of unskilled labour.

Such a fundamental change of attitude would need to stem from the governments of industrialized countries, introducing more flexibility into those regulations which control the standards to be achieved in those areas where suitable building soils are found. This change would also need to be reflected in the attitude of those organizations which finance the construction of buildings. Once these two aims have been achieved, acceptance by the general public would follow.

Jean Dethier has recently renewed his call to return to earth by the formation of the 'International Institute of Earth Architecture'. CRATerre and other authorities have pooled their efforts to create this institute of scientific research which represents the result of many years effort, building on the success of the 1981 exhibition. This ambitious project is to be built at L'Isle d'Abeau and is designed to attract the public to its museum of earth buildings. It will coordinate work being carried out in countries throughout Europe and provide expertise to the developing nations of the world. It will become a research centre for engineers, architects and others working in a multi-disciplinary manner to experiment, develop and improve earth building and repair techniques. The buildings, with a total planned area in excess of five thousand square metres will see the creation of those houses once designed by Le Corbusier, Lloyd Wright and the millionaire's mansion designed by Schindler in addition to the reconstruction of eighteenth century designs by François Cointeraux and modern designs by Hassan Fathy. The Institute aims to educate those who wish to be educated in methods of earth building and to offer information to those who request it.

The project relies on the devotion and effort of the richer countries of the world to pool their resources and expertise for the benefit of the poorer nations. An open invitation has been made to all countries to participate in research and to disseminate information. In the UK research is being undertaken into moisture content and its movement. In India, damp proof materials are being developed and tested. In New

Mexico, experiments are being made with consolidants. In Australia, water repellents are being developed but the biggest testing programme of all is taking place in France where 120 earth walls have been built to test a variety of treatments.

The results of research undertaken by the Institute are crucial and must form the basis for the conservation of existing earth buildings as well as the construction of new. We must be cautious in our repairs and take extra effort to seek, select and use traditional materials. It is essential to adopt a sympathetic and imaginative approach to conservation if the world's stock of earth buildings, including the vernacular with which we associate the heritage of this country, is to survive.

Bibliography

Albery, Jessica: 'Building in Chalk, Pisé and Cob', *Keystone* No. 4, 1962, pages 14–15.

Andrew, Martin: 'Walls With Hats On', *Country Life*, 2nd October 1986, pages 1014–5.

Ashurst, John and Nicola: *Practical Building Conservation*, Vol. 2, Gower, 1988.

Ashurst, John: *Mortars, Plasters and Renders in Conservation*, Ecclesiastical Architects and Surveyors' Association.

Bagenal, Hope: 'Building in Cob & Pisé': *Country Life*, 19th December 1947, pages 1264–5.

Barber, William: *Farm Buildings or Rural Economy. A Description of the Mode of Building in Pisé*, 1802

Baring-Gould, Rev. Sabine: *A Book of Dartmoor*, Methuen, 1923.

Barley, M. W.: *The English Farmhouse and Cottage*, Alan Sutton, 1987.

Blundell-Jones, Peter: *Mud Hut Madness*, Building Design, 9th December 1977.

Bouwens, Dirk: 'Clay Lump in South Norfolk. Observations and Recollections', *Vernacular Architecture*, Vol. 19, 1988, pages 10–18.

Brinton, Marion: *Farmhouses and Cottages of the Isle of Wight*, Isle of Wight County Council, 1987.

Brunskill, R. W.: *Traditional Farm Buildings of Britain*, Victor Gollanz, 1982.

Brunskill, R. W.: 'The Clay Houses of Cumberland', *Transactions of the Ancient Monuments Society*, New Series, Vol. 10, 1962, pages 57–80.

Burberry, Peter: *Environment and Services*, sixth edition, Mitchell, 1988.

Cherry, Bridget and Pevsner, Nikolaus: *The Buildings of England. Devon*, Penguin, 1989.

Chesher, V. M. and F. J.: *The Cornishman's House*, D. Bradford Barton Ltd, Truro, 1968.

Chudley, R.: *Building*, second edition, Longman, 1985.

Clifton-Taylor, Alec: *The Pattern of English Building*, Faber & Faber, 1977.

Clifton-Taylor, Alec: 'The Unbaked Earths. Traditional Materials of English Building', *The Connoisseur Year Book*, 1962, pages 104–109.

Cointeraux, François: Ecole d'Architecture Rurale, Premier Cahier, Paris, 1790.

Danaher, Kevin: *Ireland's Vernacular Architecture*, second edition, Cork Mercier Press for the Cultural Relations Committee of Ireland, Cork, 1978.

Davey, Norman: *A History of Building Materials*, Phoenix House, 1965.

Dethier, Jean: 'Back to Earth', *Architectural Review*, September 1990, pages 80/9–83/9.

Dethier, Jean: 'The Story of Pisé', *Architectural Review*, October 1985, pages 67/10 and 68/10.

Devon County Council: *Devon's Traditional Buildings*, 1978.

Drury, Michael and Dean, Richard: 'Rebuilding of Clay Cob Wall to Barn at Redlynch', 1987, unpublished report.

Duffy, Patrick: 'The Making of an Irish Mud Wall House', *Bealoideas*, Vol. 4, Part 4, 1933, pages 91–2.

Eaton, Ruth: 'Mud', *Architectural Review*, October 1981, pages 222–230.

Egeland, Pamela: *Cob and Thatch*, Devon Books, 1988.

Evans, E. Estyn: 'Sod and Turf Houses in Ireland', *Studies in Folk Life*, edited by Geraint Jenkins, Routledge & Kegan Paul, pages 80–90.

Fenton, A.: *Clay Building and Clay Thatch in Scotland: Studies in Folk Life*, Ulster Folk Museum, 1970, pages 28–40.

Frobisher, Howard: 'Building with Earth', *The Illustrated Carpenter and Builder*, 8th July 1949, pages 917–918.

Gailey, Alan: *Rural Houses of Northern Ireland*, John Donald, 1984.

Gailey, Alan: 'Vernacular Dwellings in Ireland', *Revue Roumaine d'Histore de l'art Serie Beaux Arts*, Vol. 13, 1976.

Gailey, Alan: 'The Housing of the Rural Poor in Nineteenth-Century Ulster', *Ulster Folklife*, part 22, 1976, pages 34–58.

Grigson, G.: *An English Farmhouse and its Neighbourhood*, Max Parish, 1948.

Handisyde, Cecil C.: *Building Materials*, Architectural Press, 1950.

Hardy, Thomas: 'Thomas Hardy on Ancient Cottages': Reproduced in *Transactions of the Association for Studies in the Conservation of Historic Buildings*, Vol. 7, 1982, page 69.

Harrison, J. R.: 'Some Clay Dabbins in Cumberland. Their Construction and Form. Part 1', *Transactions of the Ancient Monuments Society*, New Series, Vol. 33, 1989, pages 97–151.

Harrison, J. R.: 'The Mud Wall in England at the Close of the Vernacular Era', *Transactions of the Ancient Monuments Society*. New Series, Vol. 28, 1984, pages 154–174.

Holland, Henry: *Communications to the Board of Agriculture* London, 1797. (Translation of Cointeraux, François: *Ecole d'Architecture Rurale*, Premier Cahier, Paris, 1790.)

Hughes, Harold and North, Herbert L.: *The Old Cottages of Snowdonia*, 1908, republished by the Snowdonia National Park Society, 1979.

Hughes, Richard: 'Material and Structural Behaviour of Soil Constructed Walls', *Monumentum*, Vol. 26, No. 3, September 1983, pages 176–186, International Council on Monuments and Sites.

Hurst, A. E. and Goodier J. H.: *Painting and Decorating*, ninth edition, Griffin, 1980.

Innocent, C. F.: *The Development of English Building Construction*, Cambridge Technical Series, 1916.

Jaggard, W.: *Experimental Cottages. A Report on the Work of the Department at Amesbury, Wiltshire*, Department of Scientific and Agricultural Research, HMSO, 1921.

James, J. F.: 'Vernacular Architecture in the New Forest', *Annual Report of the Hampshire Field Club, New Forest Section*, 1977, pages 20–25.

Mackenzie, W. Mackay: 'Clay Castle Building in Scotland', *Proceedings of the Scottish Antiquaries Society*, Vol. 68, 1933/4, pages 117–127.

McCann, John: 'Clay and Cob Buildings', *Shire Album No. 105*, Shire Publications, 1983.

McCann, John: 'Is Clay Lump a Traditional Building Material?', *Vernacular Architecture*, Vol. 18, 1987.

McKay, W. B.: *Building Construction*, Vol. 3, fifth edition, Longman, 1974.

Meade, Martin and Garcias, Jean-Claud: 'Return to Earth', *Architecture Review*, October 1985, pages 63/10–66/10.

Mennell, Robert O.: 'Building in Chalk', *Country Life*, 10th September 1943, pages 458–460.

Mennell, Robert O.: 'Chalk Cottages', letter to *The Times*, 22nd August 1943.

Mercer, Eric: *English Vernacular Houses*, Royal Commission on Historical Monuments, England HMSO, 1975.

Mitchell, G. A. and A. M.: *Building Construction and Drawing*, Part 1, Elementary Course, Batsford, 1956.

Mulligan, Helen: 'Back to the Land', *Building Design*, 6th February 1987, pages 14–15.

Newbold, Harry Bryant: *House and Cottage Construction*, Vol. III, Caxton, 1923.

Nixon, B. H.: 'Chalk Houses': *Journal of the Junior Institution of Engineers*, Vol. 56, 1946, pages 240–243.

O'Brien, G.: *Advertisements for Ireland*, Royal Society of Antiquaries of Ireland, Dublin, 1923.

O'Danachair, Caoimhin: 'Traditional Forms of the Dwelling House in Ireland', *Journal of the Royal Society of Antiquaries of Ireland*, Vol. 102, 1972, pages 77–96.

O'Danachair, Caoimhin: 'Materials and Methods in Irish Traditional Building', *Journal of the Royal Society of Antiquaries of Ireland*, Vol. 87, 1957, pages 61–74.

Pearson, Gordon T.: 'As Cheap to Rebuild in Chalk', *Chartered Quantity Surveyor*, January 1984, page 216.

Pearson, Gordon T.: 'Chalk. Its Use as a Structural Building Material in the County of Hampshire', unpublished thesis, Architectural Association, 1982.

Pearson, Gordon T.: 'Chalk Chambers', *Traditional Homes*, June 1986, pages 30–34.

Pearson, Gordon T.: 'A Vernacular Building Style', *Celebrating Somborne* edited by Paul Marchant, The Somborne & District Society, 1989, pages 44–50.

Pearson, Gordon T.: 'Council Revives the Art of Chalk Cob', *Hampshire Magazine*, January 1984, pages 34–36.

Pearson, Gordon T.: 'Report Upon the Rebuilding of the Chalk Cob Boundary Wall to the Staff Car Park at Andover Cricklade College', Hampshire County Council, 1984, unpublished report.

Poore, G. V.: 'An experiment in Sanitation', *Country Life*, 6th July 1901.

Procter, J. M.: *East Anglian Cottages*, Providence Press, Ely, 1979.

Rees, A.: *Cyclopaedia and Universal Dictionary of Arts, Sciences and Literature*, 1819.

Reynolds, C. E. and Steadman, J. C.: *Reinforced Concrete Designers Handbook*, Viewpoint Publications for the Cement & Concrete Association, 1974.

Salzman, L. F.: *Building in England Down to 1540*, Oxford University Press, 1952.

Seaborne, M. V. J.: 'Cob Cottages in Northamptonshire', *Northamptonshire Past and Present*, Vol. 3, Part 5, 1964, pages 215–228 and Vol. 3, Part 6, 1965, pages 283–4.

Shepherd, Walter: *Flint. Its Origin, Properties and Uses*, Faber & Faber, 1972.

Smith, Peter: *Houses of the Welsh Countryside*, Royal Commission on Ancient and Historical Monuments in Wales, 1988.

Spiers, H. M.: *Technical Data on Fuel*, BNCWPC, 1962.

Sumner, Heywood: *A Guide to the New Forest*, C. Brown, Ringwood, 1923.

Taylor, R. F.: 'A Cob Dovecote at Durleigh': *Somerset Archaeological and Natural History Society*, No. 112, 1968, pages 101–103.

Tomlinson, M. J.: *Foundation Design and Construction*, Longman, 1986.

Vancouver, C.: *General View of the Agriculture of Hampshire*, Sherwood, Neely & Jones, 1813.

Walker, Bruce: 'Clay Buildings in North East Scotland', *Scottish Vernacular Buildings Working Group*, Dundee and Edinburgh, 1977.

Weeks, Charles T.: 'Method in Madness': *Building Design*, 20th January, 1978, page 19.

Whitlock, Ralph: *The Folklore of Wiltshire*, Batsford, 1976.

William, Eurwyn: *Home Made Houses*, National Museum of Wales, 1988.

Williams-Ellis, Clough: *Cottage Building in Cob, Pisé, Chalk and Clay*, Country Life, 1919.

Williams-Ellis, Clough and Eastwick-Field, John and Elizabeth: *Building in Cob, Pisé and Stabilised Earth*, Country Life, 1947.

Wright, Adela: 'Masonry Bees', *SPAB News*, Vol. 4, No. 1, January 1983.

Mud Hut News, No's 1 & 2, Devon County Council.

The Times, 11th October 1990.

Sunday Telegraph, 'Review' section, 27th January 1991.

Horncastle Standard, 18th August 1989.

Lincolnshire Echo, 28th April 1988, 11th October 1988, 24th October 1988 and 10th November 1990.

RIBA Journal, July 1990, Royal Institute of British Architects.

'The Building Bye-Laws. IX – The Winterslow Cottages', *Country Life Illustrated*, 6th April 1901, pages 430–1.

Building in Cob and Pisé de Terre, Department of Scientific and Industrial Research, HMSO, 1922.

CIBSE Guide Volume A, Design Data, The Chartered Institution of Building Services Engineers.

'Assessment of Damage in Low-Rise Buildings', *Digest 251*, August 1990, Building Research Establishment.

The Weights of Building Materials, BSS 648, 1964, British Standards Institution.

Soils for Civil Engineering Purposes, BS 1377 Part 1, 1990, and BS 5390, 1981, British Standards Institution.

The British Clayworker, 15th November 1920.

Private correspondence of the late Jessica Albery kindly provided by Helen Albery and now in the possession of the author.

Dublin Penny Journal, 1833.

Principles of Modern Building, Volume 1, third edition, Department of Scientific and Industrial Research, HMSO, 1959.

Appendices

Appendices A and B are provided to assist the reader in obtaining the specialist building materials and services referred to in the text to enable repair work to be carried out as described. The appendices do not claim to be comprehensive and are published only to provide a point of contact if goods and services are not available locally. Products mentioned in the text but not listed in Appendix A can normally be obtained from a builders merchant without difficulty.

No guarantee is given as to the quality of the materials and services offered and the purchaser should satisfy himself they are in accordance with his requirements. Not all of the companies listed are known to the author and their appearance here should not be construed as a recommendation.

APPENDIX A:
Manufacturers and suppliers of specialist building materials

	H. J. Chard & Sons	Brodie & Middleton Ltd	Rhode Design	Gunnebo Ltd	R.M.C. Industrial Minerals	W. Fein & Sons Ltd	Singleton Birch Ltd	Cathedral Works Organisation Ltd	Portmolen Paint	Rose of Jericho Ltd	Fiddles & Son Ltd	Tilcon Ltd, Somersham, Cambs.	Tilcon Ltd, Leighton Buzzard, Bucks	Hargreaves of Hull Ltd	Rory Young	Bruce & Liz Induni	Pozament Ltd	ICI Chemicals & Polymers Ltd	St Blaise Ltd	R. H. Bennett	Liquid Plastics Ltd	Unibond Ltd	J. W. Bollom	Putnams	Hüls (UK) Ltd	Snowcem PMC Ltd	Keim Mineral Paints Ltd	ARC Southern	Hargreaves Quarries Ltd	W. Aneley Ltd	Farrow & Ball
Limewater															●				●	●											
Whiting		●							●	●																					
Glue size		●							●	●																					
Clarified linseed oil		●							●											●			●								
Plaster of Paris		●							●																						
Pozament																	●			●											
Biocide wash																					●										
Quicklime	●				●		●										●												●	●	
Lime putty	●								●	●		●	●		●	●			●	●											
Casein									●											●											
Tallow									●	●										●											
HTI powder																				●											
Alum																				●											
Starch-based wallpaper paste with fungicide																						●									
Distemper									●	●																					●
Coal tar									●	●				●																	
Ethylpolysilicate																											●				
Limewash									●	●					●	●			●	●											
Silver sand									●																						
Animal hair	●					●			●						●					●											
Potassium silicate paints																											●				
Cement paint																										●					
Hydraulic lime								●																						●	
Paste pigments	●																														
Powder pigments		●							●	●	●													●							
Liquid pigments			●						●											●			●								
Oil pigments									●																						
'Gunnebo' nails				●																											

Useful names and addresses

William Aneley Ltd, Murton Way, Osbaldwick, York, PO1 3UW, telephone 0904 412624.

A.R.C. Southern, Battscombe Quarry, Cheddar, Somerset, BS27 3LR, telephone 0934 742733.

Mr R. H. Bennett, The Lime Centre, Long Barn, Morestead, Winchester, Hampshire, SO21 1LZ, telephone 0962 713636.

J. W. Bollom (Head Office) PO Box 78, Croydon Road, Beckenham, Kent, BR3 4BL, telephone 081-658 2299.

Brodie & Middleton Ltd., 68 Drury Lane, London, WC2B 5SP, telephone 071-836 3280 or 071-836 3289.

Cathedral Works Organisation (Chichester) Ltd, Terminus Road, Chichester, Sussex, PO19 2TX, telephone 0243 784225.

H. J. Chard & Sons, Feeder Road, Bristol, BS2 OTJ, telephone 0272 777681.

Farrow and Ball (Southern) Ltd, Uddens Trading Estate, Wimborne, Dorset, BH21 7NL, telephone 0202 876141, (suppliers to the National Trust).

W. Fein & Sons Ltd, Lower Mills, Holmfirth, Huddersfield, West Yorkshire, HD7 1JU, telephone 0484 682578.

Fiddles & Son Ltd, Florence Works, Brindley Road, Cardiff, CF1 7TX, telephone 0222 340323.

Gunnebo Ltd, First Floor, The Queens Suite, 1A Queens Road, Farnborough, GU14 6DJ, telephone 0252 373700.

Hardmans of Hull Ltd, Bedford Street, Kingston Upon Hull, Humberside, HU8 8AX, telephone 0482 23901.

Hargreaves Quarries Ltd, Hartly Quarries, Kirkby Stephen, Cumbria, CA17 4JJ, telephone 07683 71740.

Hüls (UK) Ltd, Edinburgh House, 43-51 Windsor Road, Slough, Berkshire, SL1 2HL, telephone 0753 71851.

ICI Chemicals & Polymers Ltd, PO Box 14, The Heath, Runcorn, Cheshire WA7 4QF, telephone 0298 768468.

Bruce & Liz Induni, Redlands, Lydeard St Lawrence, Taunton, Somerset, TA4 3SE, telephone 09847 253.

Keim Mineral Paints Ltd, Muckley Cross, Morville, Near Bridgnorth, Shropshire, WV16 4RR, telephone 074 631543.

Liquid Plastics Ltd, PO Box 7, London Road, Preston, PR1 4AJ, telephone 0722 59781.

Potmolen Paint, 27 Woodcock Industrial Estate, Warminster, Wiltshire, BA12 9DX, telephone 0985 213960.

Pozament Ltd, Swain's Park Industrial Estate, Overseal, Burton-on-Trent, Staffordshire, DE12 6JN, telephone 0283 211235, 0283 213636.

Putmans, 55 Regents Park Road, London, NW1 8XD, telephone 071-431 2935.

Rhode Design, 42 Lordship Road, London, N16, telephone 081-809 4104.

RMC Industrial Minerals, Hindlow, Buxton, Derbyshire, SK17 0EL, telephone 0298 71155.

The Rose of Jericho Ltd, Jericho Works, Deene, Nr Corby, NN17 3EJ, telephone 0780 85456.

Singleton Birch Ltd, Melton Ross Quarries, Barnetsby, South Humberside, DN38 6AE, telephone 0652 688386.

Snowcem PMC Ltd, Snowcem House, Therapia Lane, Croydon, Surrey, CR9 4RY, telephone 081-684 8936.

St Blaise Ltd, Westhill Barn, Evershot, Dorchester, Dorset, DT2 0LD, telephone 093583 662.

Tilcon Ltd, Sevenoaks Quarry, Greatness Lane, Sevenoaks, Kent, TN14 5BP, telephone 0732 453633, (Somersham (Cambrideshire) and Leighton Buzzard (Buckinghamshire) quarries only).

Unibond Ltd, Tuscam Way Industrial Estate, Camberley, Surrey, GU15 3DD, telephone 0276 685345.

Rory Young, 7 Park Street, Cirencester, Gloucestershire, GL7 2BX, telephone 0285 658826.

APPENDIX B
Equipment hire and contractors for specialist services

Pali Radici Piling – Fondedile Foundations Ltd, 192 High Street, Yiewsley, Middlesex, UB7 7BE, telephone 0895 49171.

Cementitious Grouted Anchors – Cavity Lock Systems, Factory Road, Newport, Gwent, NP9 5FA, telephone 0633 246614.

Thermal Image Hire – Livingston Hire, 2/6 Queens Road, Teddington, Middlesex, TW11 0LB, telephone 081-977 8866.

Infra-Red Thermometer Hire – The Building Services Research Information Association, Old Bracknell Lane West, Bracknell, Berks, RG12 4AH, telephone 0344 59314.

Index

Printed and bound by CPI Group (UK) Ltd, Croydon, CR0 4YY

23/10/2024

01777679-0001